Cellular Manufacturing

Mitigating Risk and Uncertainty

Industrial Innovation Series

Series Editor

Adedeji B. Badiru

Air Force Institute of Technology (AFIT) – Dayton, Ohio

PUBLISHED TITLES

Carbon Footprint Analysis: Concepts, Methods, Implementation, and Case Studies, *Matthew John Franchetti & Defne Apul*

Cellular Manufacturing: Mitigating Risk and Uncertainty, *John X. Wang*

Communication for Continuous Improvement Projects, *Tina Agustiady*

Computational Economic Analysis for Engineering and Industry, *Adedeji B. Badiru & Olufemi A. Omitaomu*

Conveyors: Applications, Selection, and Integration, *Patrick M. McGuire*

Culture and Trust in Technology-Driven Organizations, *Frances Alston*

Global Engineering: Design, Decision Making, and Communication, *Carlos Acosta, V. Jorge Leon, Charles Conrad, & Cesar O. Malave*

Handbook of Emergency Response: A Human Factors and Systems Engineering Approach, *Adedeji B. Badiru & LeeAnn Racz*

Handbook of Industrial Engineering Equations, Formulas, and Calculations, *Adedeji B. Badiru & Olufemi A. Omitaomu*

Handbook of Industrial and Systems Engineering, Second Edition *Adedeji B. Badiru*

Handbook of Military Industrial Engineering, *Adedeji B. Badiru & Marlin U. Thomas*

Industrial Control Systems: Mathematical and Statistical Models and Techniques, *Adedeji B. Badiru, Oye Ibidapo-Obe, & Babatunde J. Ayeni*

Industrial Project Management: Concepts, Tools, and Techniques, *Adedeji B. Badiru, Abidemi Badiru, & Adetokunboh Badiru*

Inventory Management: Non-Classical Views, *Mohamad Y. Jaber*

Kansei Engineering - 2-volume set
- Innovations of Kansei Engineering, *Mitsuo Nagamachi & Anitawati Mohd Lokman*
- Kansei/Affective Engineering, *Mitsuo Nagamachi*

Knowledge Discovery from Sensor Data, *Auroop R. Ganguly, João Gama, Olufemi A. Omitaomu, Mohamed Medhat Gaber, & Ranga Raju Vatsavai*

Learning Curves: Theory, Models, and Applications, *Mohamad Y. Jaber*

Managing Projects as Investments: Earned Value to Business Value, *Stephen A. Devaux*

Modern Construction: Lean Project Delivery and Integrated Practices, *Lincoln Harding Forbes & Syed M. Ahmed*

Moving from Project Management to Project Leadership: A Practical Guide to Leading Groups, *R. Camper Bull*

Project Management: Systems, Principles, and Applications, *Adedeji B. Badiru*

Project Management for the Oil and Gas Industry: A World System Approach, *Adedeji B. Badiru & Samuel O. Osisanya*

Quality Management in Construction Projects, *Abdul Razzak Rumane*

Quality Tools for Managing Construction Projects, *Abdul Razzak Rumane*

Social Responsibility: Failure Mode Effects and Analysis, *Holly Alison Duckworth & Rosemond Ann Moore*

PUBLISHED TITLES

Statistical Techniques for Project Control, *Adedeji B. Badiru & Tina Agustiady*

STEP Project Management: Guide for Science, Technology, and Engineering Projects, *Adedeji B. Badiru*

Sustainability: Utilizing Lean Six Sigma Techniques, *Tina Agustiady & Adedeji B. Badiru*

Systems Thinking: Coping with 21st Century Problems, *John Turner Boardman & Brian J. Sauser*

Techonomics: The Theory of Industrial Evolution, *H. Lee Martin*

Total Project Control: A Practitioner's Guide to Managing Projects as Investments, Second Edition, *Stephen A. Devaux*

Triple C Model of Project Management: Communication, Cooperation, Coordination, *Adedeji B. Badiru*

FORTHCOMING TITLES

3D Printing Handbook: Product Development for the Defense Industry, *Adedeji B. Badiru & Vhance V. Valencia*

Company Success in Manufacturing Organizations: A Holistic Systems Approach, *Ana M. Ferreras & Lesia L. Crumpton-Young*

Essentials of Engineering Leadership and Innovation, *Pamela McCauley-Bush & Lesia L. Crumpton-Young*

Global Manufacturing Technology Transfer: Africa-USA Strategies, Adaptations, and Management, *Adedeji B. Badiru*

Guide to Environment Safety and Health Management: Developing, Implementing, and Maintaining a Continuous Improvement Program, *Frances Alston & Emily J. Millikin*

Handbook of Construction Management: Scope, Schedule, and Cost Control, *Abdul Razzak Rumane*

Handbook of Measurements: Benchmarks for Systems Accuracy and Precision, *Adedeji B. Badiru & LeeAnn Racz*

Introduction to Industrial Engineering, Second Edition, *Avraham Shtub & Yuval Cohen*

Kansei Innovation: Practical Design Applications for Product and Service Development, *Mitsuo Nagamachi & Anitawati Mohd Lokman*

Project Management for Research: Tools and Techniques for Science and Technology, *Adedeji B. Badiru, Vhance V. Valencia & Christina Rusnock*

A Six Sigma Approach to Sustainability: Continual Improvement for Social Responsibility, *Holly Allison Duckworth & Andrea Hoffmeier Zimmerman*

Cellular Manufacturing

Mitigating Risk and Uncertainty

John X. Wang

CRC Press
Taylor & Francis Group
Boca Raton London New York

CRC Press is an imprint of the
Taylor & Francis Group, an **informa** business

CRC Press
Taylor & Francis Group
6000 Broken Sound Parkway NW, Suite 300
Boca Raton, FL 33487-2742

ISBN 13: 978-1-4665-7755-8 (hbk)

Visit the Taylor & Francis Web site at
http://www.taylorandfrancis.com

and the CRC Press Web site at
http://www.crcpress.com

To

the beautiful Grand River

by which

my sweet home

is located.

Contents

Preface

In today's business world, competitiveness defines the industrial leading edge. The drive toward maximum efficiency is constantly at the forefront of such companies' objectives. Engineering professionals across the country are striving to adopt Lean manufacturing practices to help address worries about their bottom line. Cellular manufacturing is one staple of Lean manufacturing.

Cellular manufacturing (CM) is an approach that helps build a variety of products with as little waste as possible. A cell is a group of workstations, machine tools, or equipment arranged to create a smooth flow so families of parts can be processed progressively from one workstation to another without waiting for a batch to be completed or requiring additional handling between operations. Put simply, cellular manufacturing integrates machinery and a small team of staff, directed by a team leader, so all the work on a product or part can be accomplished in the same cell, which eliminates resources that do not add value to the product.

Cellular manufacturing is a modern manufacturing system that incorporates group technology (GT) principles. Cellular manufacturing is a reorganizational manufacturing system of the traditional complex job-shop and flow-shop manufacturing systems. The efficient flow of material through a manufacturing process and its physical plant reduces material handling costs and creates a more orderly work environment. Sometimes, a group of operations can be conveniently grouped together so that raw material enters one end and finished products exit the other end. In addition, they may be arranged in a physical layout that reduces manual labor input. Such a grouping of equipment and operations is known as a "production cell." Compared to discontinuous flow through discrete operations, material flow in cells is improved. Savings opportunities arise from the elimination of downtime between operations, decreased material handling costs, decreased work-in-progress inventory and associated costs, reduced opportunity for handling errors, decreased downtime spent waiting for supplies or materials, and reduced losses from defective or obsolete products.

Decision making under risk and decision making under uncertainty are applied to cellular manufacturing, specifically in the machine cell formation step. The application works with part demand, which can be either expressed in a probability distribution (probabilistic production volume) or not expressed in probability distribution, where only the different possible values for volume that can occur are known (uncertain production volume). Decision making under risk is used to help the designer select the best cell arrangement in the case of probabilistic production volume, whereas decision making under uncertainty is used to help the designer select the best cell arrangement in the case of uncertain production volume. The objective of the design methodology has been to maximize the profit imposed by the resource capacities constraints.

I am sharing this new book with you. It is a concise yet unique reference incorporating decision making under risk into cellular manufacturing.

Author

John X. Wang, Ph.D., received a Ph.D. in reliability engineering from the University of Maryland, College Park, in 1995. He was then with GE Transportation as an engineering Six Sigma Black Belt, leading propulsion systems reliability and Design for Six Sigma (DFSS) projects while teaching GE–Gannon University's Graduate Co-Op programs and National Technological University professional short courses, and serving as a member of the IEEE Reliability Society Risk Management Committee. He has worked as a corporate Master Black Belt at Visteon Corporation, a reliability engineering manager at Whirlpool Corporation, an E6 reliability engineer at Panduit Corp., and principal systems engineer at Rockwell Collins. In 2009, Dr. Wang received an Individual Achievement Award when working as a principal systems engineer at Raytheon Company. He joined GE Aviation Systems in 2010, where he was awarded the distinguished title of principal engineer–reliability (CTH, controlled title holder) in 2013.

As a Certified Reliability Engineer certified by the American Society for Quality, Dr. Wang has authored or coauthored numerous books and papers on reliability engineering, risk engineering, engineering decision making under uncertainty, robust design and Six Sigma, Lean manufacturing, and green electronics manufacturing. He has been affiliated with the Austrian Aerospace Agency/European Space Agency, Vienna University of Technology, Swiss Federal Institute of Technology in Zurich, Paul Scherrer Institute in Switzerland, and Tsinghua University in China.

Having presented various professional short courses and seminars, Dr. Wang has performed joint research with the Delft University of Technology in the Netherlands and the Norwegian Institute of Technology.

Since his "knowledge, expertise, and scientific results are well known internationally," Dr. Wang has been invited to present at various national and international engineering events.

Dr. Wang, a CRC Press featured author, has been a top contributor of LinkedIn's Poetry Editors & Poets group. He has contributed to the discussions including:

- Writing a sonnet is notoriously difficult due to the strict pentameter and rhyming pattern; does anyone prefer/enjoy writing this form of poetry?
- Do you proceed by images or by words when you write?

Connections
- CRC Press Featured Author: http://www.crcpress.com/authors/i351-john-wang
- LinkedIn: http://www.linkedin.com/pub/john-wang/3/2b7/140

chapter one

The handling of bottleneck machines and parts in cellular manufacturing

Cellular manufacturing is a reorganizational manufacturing system of the traditional complex job-shop and flow-shop manufacturing systems. The purpose of this chapter is to present a survey of the current state of existing methods that can best be used in the handling of the bottleneck machines and parts problem, which results from the cellular manufacturing system design.

1.1 Cellular manufacturing: Workplace design based on work cells

Cellular manufacturing (CM) is a model for workplace design, and has become an integral part of Lean manufacturing systems. Cellular manufacturing is a modern manufacturing system incorporating group technology (GT) principles. Group technology is a manufacturing philosophy in which the parts having similarities (geometry, manufacturing process, or function) are grouped together to achieve a higher level of integration between the design and manufacturing functions of a firm. Group technology or family-of-parts concepts are often used in cellular design. Group technology is the process of studying a large population of different workpieces and then dividing them into groups of items that have similar characteristics. The process may be executed with the aid of a computer, and has been used to divide parts into groups for use with computer-aided design (CAD)/computer-aided manufacturing (CAM) processing. Family of parts is the process of grouping workpieces into logical families so that they can be produced by the same group of machines, tooling, and people with only minor changes to procedures or setup.

Cellular manufacturing is a reorganizational manufacturing system of the traditional complex job-shop and flow-shop manufacturing systems. In cellular manufacturing, production workstations and equipment are arranged in a sequence that supports a smooth flow of materials and components through the production process with minimal transport or delay. Implementation of this Lean method often represents the first major shift in production activity, and it is the key enabler of increased production velocity and flexibility, as well as the reduction of capital requirements.

Cellular manufacturing refers to a manufacturing system wherein the equipment and workstations are arranged in an efficient sequence that allows a continuous and smooth movement of inventories and materials to produce products from start to finish in a single process flow, while incurring minimal transport or waiting time, or any delay for that matter. Rather than processing multiple parts before sending them on to the next machine or process step (as is the case in batch-and-queue or large-lot production), cellular manufacturing aims to move products through the manufacturing process one piece at a time, at a rate determined by customers' needs. Cellular manufacturing can also provide companies with the flexibility to vary product type or features on the production line in response to specific customer demands. The approach seeks to minimize the time it takes for a single product to flow through the entire production process.

Cellular manufacturing and work cells are at the heart of Lean manufacturing. Their benefits are many and varied. They increase productivity and quality. A work cell is a group of workstations, machines, or equipment arranged such that a product can be processed progressively from one workstation to another without having to wait for a batch to be completed or requiring additional handling between operations. Work cells may be dedicated to a process, a subcomponent, or an entire product. Work cells are conducive to single-piece and one-touch manufacturing methods, and are often found as part of Lean manufacturing applications. Work cells may be designed for the office as well as the factory.

Work cells simplify material flow, management, and even accounting systems. This one-piece flow method includes specific analytical techniques for assessing current operations and designing a new cell-based manufacturing layout that will shorten cycle times and changeover times. To make the cellular design work, an organization must often replace large, high-volume production machines with small, flexible, "right-sized" machines to fit well in the cell. Equipment often must be modified to stop and signal when a cycle is complete or when problems occur, using a technique called autonomation (or jidoka).

In summary, cellular manufacturing is a manufacturing process that produces families of parts within a single line or cell of machines operated by machinists who only work within the line or cell. A cell is a small-scale, clearly defined production unit within a larger factory. This unit has complete responsibility for producing a family of like parts or a product. All of the necessary machines and manpower are contained within this cell, thus giving it a degree of operational autonomy. Each worker is expected to have mastered a full range of operating skills required by his or her cell. Therefore, systematic job rotation and training are necessary conditions for effective cell development. Complete worker training is needed to ensure that flexible worker assignments can be fulfilled. The purpose of this chapter is to present a survey of the current state of existing methods that can best be used in the handling of the bottleneck machines and parts problem, which results from the cellular manufacturing system design.

1.2 *How to establish cellular manufacturing systems*

A cellular manufacturing system (CMS) offers a system approach to the reorganization of the inflexible, repetitive batch of job shop and mass production of flow shop to more flexible small-lot production. CMS is a production system that will allow a set of dissimilar machines to be grouped into a machine cell to process a group of a product/part family. A product/part family is a group of parts that can be produced by the same sequence of machining operations because of a similarity in design and processing attributes. This transformation often shifts worker responsibilities from watching a single machine to managing multiple machines in a production cell. Although plant floor workers may need to feed or unload pieces at the beginning or end of the process sequence, they are generally freed to focus on implementing total productive maintenance (TPM) and process improvements. TPM seeks to engage all levels and functions in an organization to maximize the overall effectiveness of production equipment. This method further tunes up existing processes and equipment by reducing mistakes and accidents. Whereas maintenance departments are the traditional center of preventive maintenance programs, TPM seeks to involve workers in all departments and levels, from the plant floor to senior executives, to ensure effective equipment operation. Integrating CMS and TPM, production capacity can be incrementally increased or decreased by adding or removing production cells.

The main objective of CMS is to achieve benefits and efficiencies in manufacturing. The benefits and efficiencies of CMS include:

- Work-in-process (WIP) reduction
- Lead time or throughput time reduction
- Productivity improvement
- Quality improvement
- Better scheduling
- Simplicity in tool control
- Enhanced flexibility and visibility
- Better teamwork and communication

The major focus of CMS is the cell formation (CF). Cellular manufacturing requires a fundamental paradigm shift from "batch and queue" mass production to production systems based on a product-aligned "one-piece flow, pull production" system. Batch and queue systems involve the mass production of large inventories in advance, where each functional department is designed to minimize the marginal unit cost through large production runs of similar products with minimal tooling changes. Batch and queue entail the use of large machines, large production volumes, and long production runs.

CMS also requires companies to produce products based on potential or predicted customer demands, rather than actual demand, due to the lag time associated with producing goods by batch and queue functional department. In many instances this system can be highly inefficient and wasteful. Primarily, this is due to substantial work-in-process, or WIP, being placed on hold while other functional departments complete their units, as well as the carrying costs and building space associated with built-up WIP on the factory floor. In a batch-and-queue system, the production flow begins with a large batch of units from the parts supplier. The parts make their way through the various functional departments in large "lots," until the assembled products are eventually shipped to the customer. The following steps and techniques are commonly used to implement the conversion to cellular manufacturing.

Step 1: Understanding the current conditions. The first step in converting a work area into a manufacturing cell is to assess the current work area conditions, starting with product and process data. For example, PQ (product type/quantity) analysis is used to assess the current product mix. Organizations also typically document the layout and flow of the current processes using process route analyses and value stream mapping (or process mapping).

Step 2: Measuring time elements. The next activity is often to measure time elements, including the cycle time for each operation and the lead time required to transport WIP between operations. Takt time, or the number of units each operation can produce in a given time, is another important time element to assess. Time elements are typically recorded on worksheets that graphically display the relationship between manual work time, machine work time, and operator movement time for each step in an operation. These worksheets provide a baseline for measuring performance under a cellular flow.

Step 3: Converting to a process-based layout. The next step involves converting the production area to a cellular layout by rearranging the process elements so that processing steps of different types are conducted immediately adjacent to each other. For example, machines are usually placed in a U or C shape to decrease the operator's movement, and they are placed close together with room for only a minimal quantity of WIP. The process flow is often counterclockwise to maximize right-hand maneuvers of operators.

Step 4: Continuously improving the process. This step involves fine-tuning all aspects of cell operation to further improve production time, quality, and costs. Kaizen, TPM, and Six Sigma are commonly used as continuous improvement tools for reducing equipment-related losses such as downtime, speed reduction, and defects by stabilizing and improving equipment conditions. In some cases, organizations seek to pursue a more systemic redesign of a production process to make a "quantum leap" with regard to production efficiencies and performance. Production Preparation Process (3P) is increasingly used as a method to achieve such improvements.

To enable a smooth conversion, it is typically necessary to evaluate the machines, equipment, and workstations for movability and adaptability, and then develop a conversion plan. In many cases, it is helpful to mock up a single manufacturing cell to assess its feasibility and performance.

1.3 Forming cells to capture the inherent advantages of group technology (GT)

The GT approach is an original philosophy that exploits the proximity among the attributes of given objects. Cellular manufacturing is an application of GT in manufacturing. CM involves processing a collection of similar parts (part families) on a dedicated cluster of machines or manufacturing processes (cells). The cell formation problem in cellular manufacturing systems is the decomposition of the manufacturing systems into cells. Part families are identified such that they are fully processed within a machine group. The cells are formed to capture the inherent advantages of GT such as

- Reduced setup times
- Reduced in-process inventories
- Improved product quality
- Shorter lead time
- Reduced tool requirements
- Improved productivity
- Better overall control of operations

Facing intensified competition in a growing global market, manufacturing companies are reengineering their integrated production systems to achieve Lean manufacturing. In recent years, there appears to be a trend showing increasing popularity of CM to achieve this goal. Substantial research has been performed to improve the grouping of machines and parts into cells as a result of this trend toward CM.

The conversion from traditional job-shop production to CM brings a new culture context to the worker team. In creating cells, workers with process-oriented skills must be divided into part-oriented teams and assigned to cells with heterogeneous processes. Worker training becomes an integral part of cellular team formation and success. In creating empowered teams, additional technical, teamwork, and administrative skills must be developed among the workforce. Cell productivity depends not only on the technical and administrative skills that the workers possess but also the effective interaction among team members.

Given an existing labor pool, it is desired to extract multiple teams. This would be required if we were to shift from a noncellular manufacturing environment to a cellular manufacturing environment. In this case, we would need to determine which skilled individuals to place together in which cell. Multiple teams may comprise the entire existing labor pool or just some subset of that pool. For example, if an entire segment of an organization were shifting to cellular manufacturing, the entire labor pool might need to be redistributed. However, if only a portion of the organization was being formed into a small number of teams, the entire labor pool would be considered but only a portion of it allocated. Depending on the nature of the work, cells could have the same makeup or vary from cell to cell. We assume the labor pool itself is segregated into skill categories. Each member of the labor pool is assigned to one and only one skill category. The categories are defined according to the jobs or roles that need to be fulfilled on the team(s). For example, a team may require a milling machine operator, a turning machine operator, an inspector, and an assembler. Each of these would become a skill category and any individuals belonging to the labor pool would have to be classified according to one of these skills. In some instances, workers will receive additional cross-training to become multifunctional, but we assume that initially they have a key skill or experience. Furthermore, even cells with extensive cross-training will benefit from having lead individuals for each task to assist in problem solving and training others.

The problem with cell design is that it is a very complex exercise with wide-ranging implications. Normally, cell design is understood as the problem of identifying a set of part types that are suitable for manufacture on a group of machines. However, there are a number of other strategic level issues, such as level of machine flexibility, cell layout, type of material handling equipment, types and number of tools and fixtures, and so on, that should be considered as part of the cell design problem. Any meaningful cell design must be compatible with the tactical/operational goals, such as high production rate, low WIP, low queue length at each workstation, and high machine utilization.

Manufacturing efficiencies have gained by constructing systems from independently designed and optimized tasks. Recent theories and practice have extolled the virtues of team-based practices that rely on human flexibility and empowerment to improve integrated system performance. The formation of teams requires considering innate tendencies and interpersonal skills as well as technical skills.

1.4 The cell formation problem

The major focus of CMS is the cell formation (CF). It is the ability to classify parts into part families and forming machines into machine cells that are dedicated to the manufacture of a specific part family. Unfortunately, the grouping is not always possible to ensure that all parts of a family can be processed within a machine cell and tend to create the problem of the bottleneck machines and parts issues in the CF effort.

Cell formation is a major step in cellular manufacturing planning and implementation. Cell formation is a process that provides information about work cells. Given production information about parts and machines as input, the output of cell formation is the number of work cells and the parts and machines inside each work cell. There have been many cell formation grouping methodologies since 1970. In order to achieve higher productivity from cellular manufacturing, certain design goals such as process completeness and resource usage should be established in the cell formation problem. In addition, the cell formation problem is controlled by managerial and technical constraints. Examples include the number of possible new machines, separation of hazardous processes from other activities, work cell size, and the number of operators/equipment assignable to a work cell. Cell formation design goals establish the criteria for the quality of cell formation output.

During the last three decades, there has been a significant amount of research on cell formation and many solution procedures have been developed based on different methodologies for cell formation. During this period, the cell formation problem has evolved from the basic machine–part problem to more complicated problems considering multiple objectives. At first, the focus of researchers was on the basic machine–part cell formation problem that only considers the machines a part visits regardless of other production information. With progress in cell formation methodologies, the cell formation problem is becoming more practical with the inclusion of more production information. Examples include considering the information about the order in which a part visits machines (production sequence) or different ways of making a part by using existing equipment (alternative routing). The leading methodologies to grouping machines into machine cells and parts into part families are as follows:

- Machine-component group analysis (McCormick et al., 1972)
- Coding and classifications (Dunlap and Hirlenger, 1983; Chandrasekaran and Rajagopalan, 1986)
- Similarity coefficients analysis (McAuley, 1972; Gupta and Seifoddini, 1990)
- Knowledge-based (ElMaraghy and Gu, 1988; Ang et al., 1994)
- Fuzzy theory with similarity coefficient and mathematical programming (Güngör and Arikan, 2000)
- Heuristics and algorithms (Logendran, 1990; Ang and Hegji, 1997)
- Multicriteria (Mansouri, Mooattar, and Newman, 2000)

The comprehensive reviews for the cell formation design problem and the solution approaches can be found in a review by Singh (1993).

1.5 The cell formation problem

As part of the company's growth plan, Grand Cycles (GC) is trying to introduce a new line of mountain bikes. Currently, the facility produces regular bikes for the domestic market. It can produce these bikes at a rate of 200 bikes per day (50,000 bikes per year). The demand for both regular and mountain bikes for the last 5 years has been steadily increasing. Presently, GC imports mountain bikes to meet the demand. This is a costly option. As the industrial engineering team for GC, we will analyze the impact of producing enough mountain bikes and regular bikes to meet the demand for the next 5 years, suggest a better material handling option, and investigate staffing within the organization.

1.5.1 Demand profile

To determine what production levels need to be met, the company's demand data can be analyzed and used to forecast future demand. The demand data for the past 5 years can be seen in Table 1.1. Using these figures for demand, it is possible to forecast future demand of the bicycles.

These market demand figures are placed into Microsoft Excel and a regression analysis is performed to predict the demand for the next 5 years. This analysis can be seen in Figure 1.1. Regression analysis is widely used for prediction and forecasting, where its use has substantial overlap with the field of machine learning. Regression analysis is also used to understand which among the independent variables are related to the dependent variable, and to explore the form of these relationships. In

Table 1.1 Combined Demands 2010–2014

Year	Yearly demand	Daily demand (250 working days per year)
2010	75,000	300
2011	82,000	328
2012	80,000	320
2013	77,000	308
2014	79,000	316

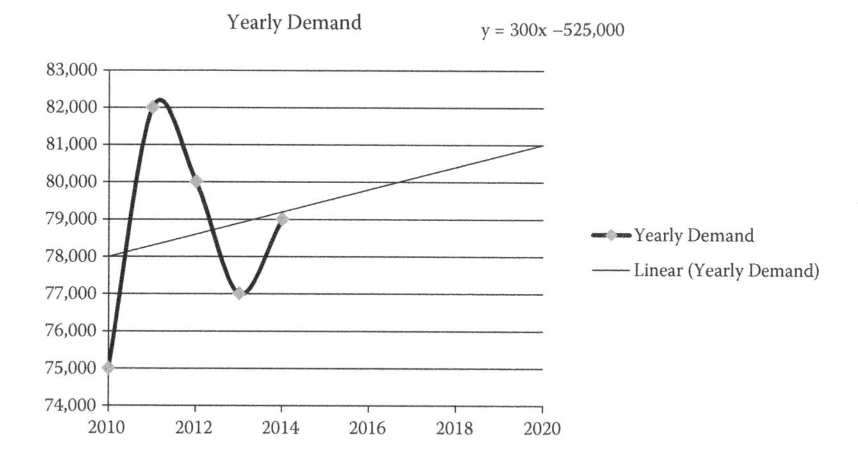

Figure 1.1 Regression line for future bicycle demand.

restricted circumstances, regression analysis can be used to infer causal relationships between the independent and dependent variables.

Regression analysis is a statistical tool for the investigation of relationships between variables. Usually, the investigator seeks to ascertain the causal effect of one variable upon another, for example, the effect of a price increase upon demand or the effect of changes in the money supply upon the inflation rate. To explore such issues, the investigator assembles data on the underlying variables of interest and employs regression to estimate the quantitative effect of the causal variables upon the variable that they influence. The investigator also typically assesses the "statistical significance" of the estimated relationships, that is, the degree of confidence that the true relationship is close to the estimated relationship.

Based on regression analysis, the following prediction is used to analyze the production needs of GC (Table 1.2). A regression analysis of the past market data showed that by the year 2020, GC will need to produce 324 bikes per day (81,000 bikes per year). These demand predictions will allow the engineering team to plan production levels for the next 6 years.

Table 1.2 Predicted Combined Demands 2014–2020

Year	Predicted demand	Predicted daily demand (250 working days)
2015	79,500	318
2016	79,800	319
2017	80,100	320
2018	80,400	322
2019	80,700	323
2020	81,000	324

1.5.2 Assumptions

- Production at 82,000 units would be interpreted as 328 bikes a production day.
  - Of those 328 bikes, 200 would be regular bikes and 128 would be mountain bikes.
- 250 workdays/year; 2,080 work hours/year; 8-hour workdays.
- Anything not produced is subcontracted or purchased from the market for the final assembly.
- In the proposed cellular model, decreasing process time is the equivalent of adding machines/capacity.
- The processing of handlebars for each bike takes a similar amount of time due to restrictions of the software.
- A 100-hour warm-up period was used to reach steady state.

1.5.3 Support of assumptions about key model parameters

After doing research online and by looking in the assembly instruction book, our recommended travel speed for forklifts and autoguided vehicles is 5 mph (440 fpm). This is calculated by taking many factors into consideration such as accident prevention, the type and size of load being carried, the amount of traffic on the vehicle's path, and the manufacturer's recommended speed of the vehicle.

The operation time is set up to simulate an 8-hour shift with a 100-hour warm-up period. This was designed as a one-shift-only manufacturing facility because the total demand for bikes is 328 per day, which breaks down into 200 regular bikes and 128 mountain bikes. The way our simulation model is set up is one 8-hour shift produces the 200 demanded regular bikes with a small amount of surplus and the 128 demanded mountain bikes. We simulated in our model that the demand would increase over time so we simulated the effects of adding a second shift.

Table 1.3 Data for Current Situation Analysis

	Original facility (regular bikes only)			
	Time/Part/ Machine	Number of machines	Number of parts per bike	Total time per bike (sec)
Cutting	15	2	10	75
Bending	30	2	3	45
Welding	60	8	6	45
Casting	60	1	5	300
Molding	45	2	2	45
Forging (large)	60	1	1	60
Forging (small)	30	1	2	60
Drill press	20	1	0	0
Punch press	30	1	0	0

1.5.4 *Grand Cycle's current (as-is) model*

A current model of Grand Cycle's (GC) facility is produced to form a baseline of production based on data from GC. This data can be seen in Table 1.3. This data is taken from GC and is used to build the model of the company's current situation. The as-is model layout is that of a job-shop facility; the raw material enters the system and proceeds to transfer to whichever machine is needed. The material is continuously routed back and forth through the machines for different operations. The current production is 200 regular bikes per day and no mountain bikes built in the facility. There is a high quantity of casting and final assembly machines in the system to account for the longer operation times. Bottlenecks were created because the machines were set up as first in, first out, so if all the frames went to casting then the seats and handlebars would have to wait. It can be seen in the model layout in Figure 1.2 that the workspace is used very inefficiently. This will change when we modify the model for the cellular manufacturing system.

1.5.5 *A flexible manufacturing system*

The entire company is modeled around the exact concept of a flexible manufacturing system (FMS). GC's goal is to produce reasonably priced customized products of high quality that can be delivered quickly to customers. GC has the ability to produce different parts without major retooling, can convert processes between new and old models of products (regular and mountain bikes), and can offer a wide variety of customizations to its products for customers.

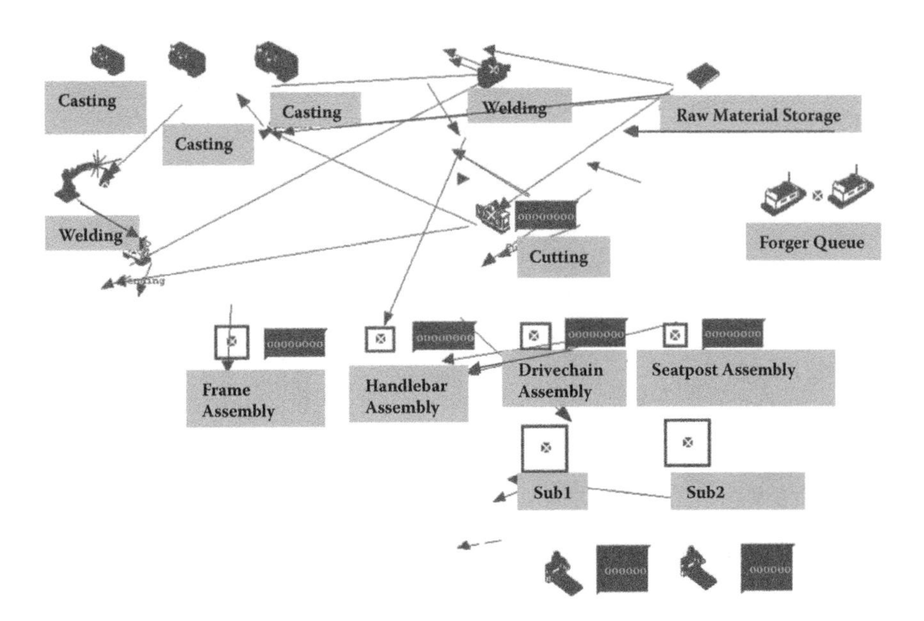

Figure 1.2 GC's current (as-is) model.

This is achieved by making the main parts in-house with machines that are semiautomated and very flexible, and subcontracting the customized parts to vendors or purchasing them from the market. By keeping the machines semiautomated they can be easily retooled for variations or new models.

Our model incorporates FMS by having our machines semiautomated thus allowing quick changeovers but keeping quality high with automated control. By using automated guided vehicles (AGVs) to transport materials between raw material receiving/storage, a lower direct labor cost is achieved.

The first task was to model GC as it is in its current state, a job shop. By being a job shop, the cycle company can focus on high quality and customization. This became a problem because as the demand increases, GC is starting to incur problems with meeting that demand. The goal of our project is to model GC as a cellular manufacturing facility and study the benefits, if any, that this layout has to offer, such as increased capacity and reduced costs.

Cellular manufacturing still offers a high level of flexibility in manufacturing. Similar processes are grouped into cells. The work is then divided into categories and each cell is assigned a category of work. Currently, the cellular manufacturing model that is built has a severe bottleneck in it: the bike frame cell. This cell is the main bottleneck in the operation because of the high number of operations performed to build a bike frame. The bottlenecks in the system were alleviated by increasing the number of frame and handlebar cells. This bottleneck was making the cellular manufacturing layout extremely inefficient and actually unfavorable when compared to the original job shop.

1.5.6 A deterministic model to estimate system performance

A deterministic model is used to estimate the system performance. A bottleneck model is a mathematical description of a deterministic model. The bottleneck model is used to provide starting estimates of the FMS design parameters, such as the production rate and the number of workstations. We performed this bottleneck model with GC's data. We performed the bottleneck model assuming that 333 bikes will be leaving the system.

Cell formation design is obviously a key issue in CMS design. In general, for a production facility with a given number of machines and part mix to be processed in the facility, there are three specific decisions in cell formation design:

1. Number of manufacturing cells to be established
2. Machines constituting each cell
3. Parts assigned to each cell

As shown in Table 1.4, part mix is equal to total parts produced by cell divided by total parts produced by system, that is,

Part mix = Total parts produced by cell/Total parts produced by system

Customized products, shorter product life cycles, and unpredictable patterns of demand have challenged the manufacturers to improve the efficiency and productivity of their production activities. Manufacturing systems should be able to adjust or respond quickly to adopt necessary changes in product design and product demand without major investment. Since shorter product life cycles are an increasingly important issue

Table 1.4 Part Mix Is Equal to Total Parts Produced by the Cell Divided by the Total Parts Produced by the System

Cell	Parts	Part mix
Frame cell	333	0.2766
Handlebar cell	333	0.2766
Seatpost cell	333	0.2766
Drivechain cell	205	0.1703
Total	1204	1

in cellular manufacturing, one cannot assume that the designed cells will remain effective for a long time. Ignoring the planned new product introductions would necessitate subsequent ad hoc changes to the CMS causing production disruptions and unplanned costs. Thus, one has to incorporate the product life cycle changes in the design of cells. This type of model is called the multiperiod CMS or dynamic CMS. Since the formed cells in the current period may not be optimal for the next period, the reconfiguration of the cells is required.

By completing the calculations to determine the part mix (Table 1.4), the bottleneck is assessed as follows (see Table 1.5):

- By using the average workload and workload per server for each cell we were able to identify the bottleneck as the frame cell with a ratio of 50.82.
- Once the bottleneck was identified the next calculation to complete was the maximum production rate produced by the system; this was determined to be 70.8 (pieces/hour).
- Other calculations performed during this model consist of the production rate, utilization, and mean busy servers for each cell.

The bottleneck identification and assessment are performed as follows:

WLi = ΣΣ [Processing time (ijk) × Operation frequency (ijk) × Part mix fraction]
Bottleneck station has largest WLi/si ratio
Production rate of each product = Rp* × Part mix fraction
Utilization of each station = Server workload/Bottleneck server workload
Mean number of busy server = Station workload × (Rp*)

When comparing the data found by performing the bottleneck model and the data obtained from running the simulation (Table 1.6), the outcomes were consistent with each other. From the bottleneck model it can be suggested that the frame cell is the bottleneck of the operation; when looking at the simulation output it can be seen that the frame cell, handlebar cell, and seat cell are all utilized 100% of the time. By adding two additional frame cells, the utilization stays at 100% but the number of total entries into the system increases from 100 to 336, just enough to complete the 333 bicycles.

When the additional frame cells were added, it caused the handlebar cell to be the bottleneck, holding up continued operations. To fix this problem, an additional handlebar cell was also added and the utilization stayed steady at 100%, but the number of entries increased from 282 to 564, which created enough to complete the 333 bikes. These increases in utilization support the fact that they are both bottlenecks in the system.

Table 1.5 Bottleneck Identification and Assessment

Part	Part mix	Freq.	Process time (sec)	Average workload (WL)	# Servers at station (Si)	WL/Si	Rp^* (pc/sec)	Rp^* (pc/hr)	Production rate (pc/hr)	Utilization	Mean busy servers
Frame cell	0.277	1	735	203.285	4	50.82	0.02	70.8	19.5918	100.00%	1.0000
Handlebar cell	0.277	1	165	45.635	4	11.41			19.5918	22.45%	0.8980
Seatpost cell	0.277	1	60	16.595	2	8.297			19.5918	16.33%	0.3265
Drivechain cell	0.170	1	150	25.540	1	25.54			12.0610	50.25%	0.5025

Table 1.6 Output Report for Suggested Production System

```
- - - - - - - - - - - - - - - - - - - - -
General Report
- - - - - - - - - - - - - - - - - - - - -
Scenario        : Normal Run
Replication     : 1 of 1
Warm-Up Time    : 100 hr
Simulation Time : 108 hr
- - - - - - - - - - - - - - - - - - - - -

LOCATIONS
```

Location Name	Scheduled Hours	Capacity	Total Entries	Average Minutes Per Entry	Average Contents	Maximum Contents	Current Contents	% Util
Raw materials	8	999999	2896	0.0	0	1	0	0.0
Frame cell.1	8	1	112	4.28	1	1	1	100.00
Frame cell.2	8	1	112	4.28	1	1	1	100.00
Frame cell.3	8	1	112	4.28	1	1	1	100.00
Frame cell	24	3	336	4.28	1	3	3	100.00
Handlebar cell	8	2	564	1.70	2	2	2	100.00
Seat cell	8	1	365	1.31	1	1	1	100.00
Drive cell	8	1	496	0.83	0.86	1	1	86.06
Subassembly 1	8	999999	732	317.08	483.55	582	526	0.05
Subassembly 2	8	999999	334	0.0	0	1	0	0.0
Drive cell queue	8	999999	600	75.43	94.29	199	104	0.01
Seat cell queue	8	999999	5684	431.50	5109.78	5357	5320	0.51
Handlebar cell queue	8	999999	3704	390.35	3012.25	3234	3142	0.30
Frame cell queue	8	999999	6470	437.35	5895.16	6187	6137	0.59
Final assembly	8	999999	353	28.32	20.82	27	20	0.0
Seat sub 1	8	999999	639	244.02	324.85	386	311	0.03
Frame sub 1	8	999999	352	28.40	20.82	27	18	0.0
Handlebar sub	8	999999	3348	416.40	2904.42	3016	3014	0.29
Drive sub	8	999999	3295	435.77	2991.42	3095	3095	0.30

Because the seat cell is receiving 365 entries and meets the required quantity with one cell, it is sufficient and does not need another. Simulation output used for comparison to the bottleneck model is shown in Table 1.7.

CM is a proven technique in batch-type manufacturing. It improves the efficiency of manufacturing systems through a reduction in setup times, in-process inventories, and throughput times. This has been shown in numerous successful cases of CM implementation. Despite this, some studies suggest that the conversion from conventional batch-type manufacturing to cellular manufacturing may not always be beneficial. This is mostly due to the imbalanced workload on machines in machine cells, which leads to the accumulation of inventories in front of bottleneck machines. Imbalanced workloads also cause underutilization of non-bottleneck machines. In addition, the organization of machines into dedicated machine cells decreases the flexibility in machine selection for the processing of parts.

1.5.7 Develop the future state: To-be model

After construction of the as-is models, the design team began to construct a to-be model that represented proposed modifications and goals for the selected business processes (Figure 1.3). One of the principal goals set by the project team was the development of a scheduling and operations methodology that would reduce lead times and WIP inventory levels. To accomplish these goals, the organization needed a system that would provide the capability to produce the right product at the right time for the right customer. This would require reengineering several aspects of the selected business processes that included the elimination of long established shop floor scheduling, performance measurement, and production methodologies.

Due to the limitations of the student model, it became necessary to black box each cell and compute the processing times for each entity at each process. Each cell is responsible for producing either a frame, seat assembly, drive assembly, or handlebar assembly. The raw materials are brought from storage to their respectable queues. They are then pushed into their cells where the process is performed. After processing they are moved into subassembly stations. The handlebars and frames are joined together in subassembly 1, and the drive assembly and seat assembly are joined in subassembly 2. The two subassemblies are then joined together at the final assembly station into either a mountain bike or regular bike. This was a very difficult model to try and balance because it takes only 37 seconds for processing in the seat cell, whereas it takes almost 5 minutes to process a frame.

Creativity in the to-be model was built into the system in a number of ways. The locations were built differently in the as-is model compared to

Table 1.7 Simulation Output Used for Comparison to the Bottleneck Model

ENTITY ACTIVITY

Entity Name	Total Exits	Current Quantity in System	Average Minutes in System	Average Minutes in Move Logic	Average Minutes Wait for Res, etc.	Average Minutes in Operation	Average Minutes Blocked
Regular frame	0	3746	-	-	-	-	-
Mountain frame	0	2432	-	-	-	-	-
Regular seat	206	3726	3548.48	0.0	0.0	30.62	3517.85
Mountain seat	128	2432	3687.10	0.0	0.0	2.16	3684.94
Regular handlebar	334	6158	3601.64	0.0	0.0	31.70	3569.93
Drivechain	200	3200	3128.20	0.0	0.0	30.83	3097.36
Regular bike	205	0	3551.62	0.0	0.0	64.83	3486.78
Mountain bike	128	0	3716.95	0.0	0.0	63.50	3653.45

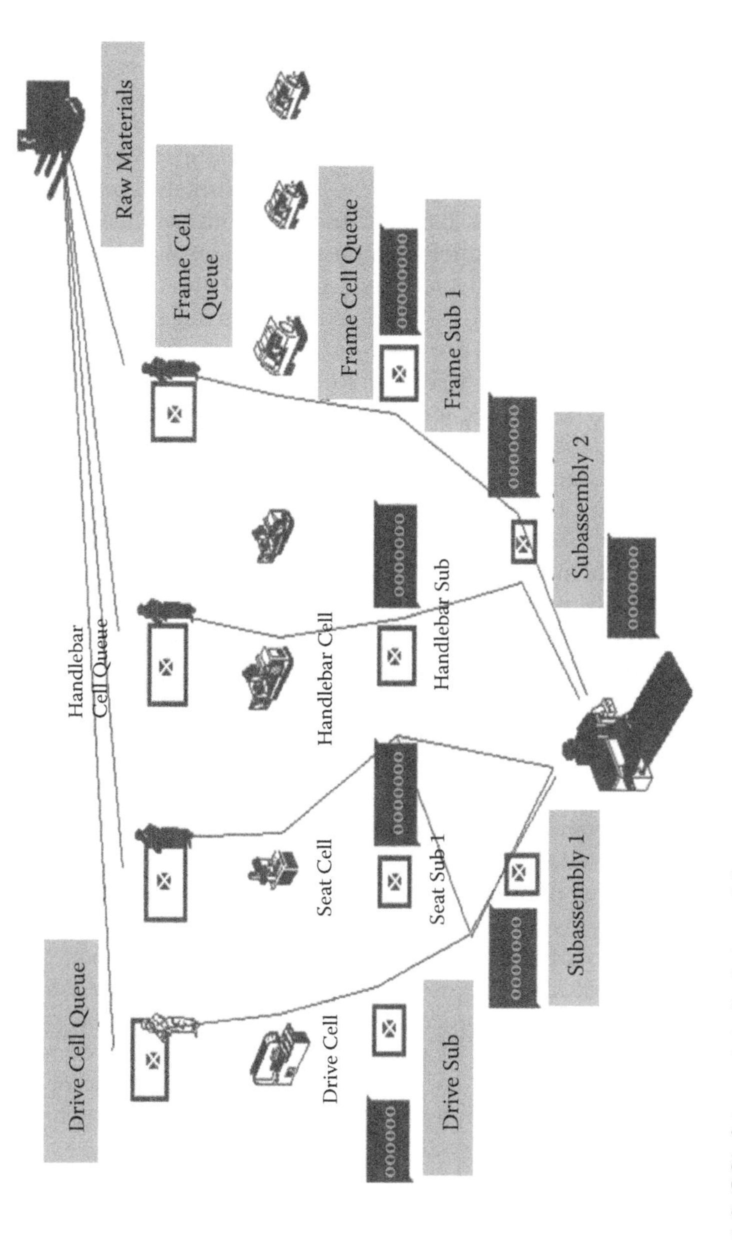

Figure 1.3 GC's future state (to-be) model.

the to-be model. In the as-is model the structure of the layout is that of a job shop, with limited capacities for each process. This model also breaks out each process by function, which sets constraints and limitations in the system. In the simulation, bottlenecks are spotted in the amount of time that entities are blocked trying to get to the next process.

The to-be model is structured like a cellular manufacturing system, using group technology and integrated systems. This model is flexible and promotes Lean concepts, such as reducing waste and inventory. The to-be model has the capacities for each of the cells set infinitely. For this model, rather than having processes blocked, the problems would arise when inventory levels would be higher than desired. This is a creative simulating method since it does not depict a realistic model because of the infinite capacity levels. In addition, AGVs correlated with the rate of arrivals to supply the beginning stages of the processing system. The number of fork trucks was set to allow the flow of arrivals into the system. This thinking was designed to avoid a bottleneck at the beginning of the model. Realistically, with smaller parts, such as bike seats and drivechains, conveyors would be used to avoid the large costs of automated storage and retrieval systems.

People are used to simulating the activities that take place in work cells. The cellular manufacturing layout removes the boundaries of the job shop by only being responsible for specific tasks. The cellular layout also uses top-down modeling, having arrivals flowing into the top, and having finished units exiting from the bottom of the model with machinists following the work throughout the system in designated cells. Two work shifts per day were built into the model. Using the philosophy of a flexible manufacturing system, the beginning of these shifts can be called to start at different times and for different durations depending on customer product demand. As the case study showed there was a definite fluctuation in demand and as a result supply should try to match that demand and minimize inventory.

One of the most popular methods for reducing inventory levels and customer lead times is cellular manufacturing. It provides a manufacturer with the ability to streamline and combine the isolated functions that make up a manufacturing process and eliminates non-value-added activities, the essence of business process reengineering. For example, material handling and changeover time can be reduced by grouping products into production families that can be produced across a single line or cell of machines. Many who have implemented cellular manufacturing have significantly reduced inventory and customer lead times.

1.5.8 Results and recommendations

While completing the simulations we found that the bottlenecks in the system were primarily the frame cell and the handlebar cell. To fix these we decided to add two frame cells and one handlebar cell. By making these additions the output was increased enough to cover the projected demand; we were able to output 205 regular bikes and 128 mountain bikes. We were able to reduce labor by adding AGVs to our system to transport raw materials from the storage area to the appropriate cells. Also we reduced the number of workers needed by incorporating a floating worker design. We recommended implementing the group technology cellular manufacturing layout, adding additional frame and handlebar cells, and using the flexible workers and the AGVs for the transport of raw materials.

As compared to cell formation research, the study of handling bottleneck machines and parts problems are much less. This study presents a concise review of the current state of existing literature on the handling of bottleneck machines and parts aspects of cell formation in the design of CMS. The advantages and disadvantages of solution methods are discussed and contrasted.

1.6 Lean manufacturing

Methods to solve the bottleneck machine and parts problems in CMS all have benefits and can make the production environment a more efficient, more profitable arena. This research has found the methods developed to date to handle bottleneck machines and parts in cellular manufacturing systems include the following:

- Algorithms
- Mathematical programming
- Heuristic rules
- Duplication
- Subcontracting
- Modifying the process plan

The research confirms that the methods for handling bottleneck machine and parts problems involve both quantitative and qualitative approaches. Efficiency among the methods varies, and there is no complete and universally satisfactory method to date that can eliminate bottleneck parts problems in cellular manufacturing systems.

Lean manufacturing is a business model and a collection of tactical methods that emphasize eliminating non-value-added activities (waste) while delivering quality products on time with the least amount of cost and greater efficiency. In the United States, Lean implementation is rapidly expanding throughout diverse manufacturing and service sectors such as the aerospace, automotive, electronics, furniture production, and health care industries as a core business strategy to create a competitive advantage.

While the focus of Lean manufacturing is on driving rapid, continual improvement in cost, quality, service, and delivery, significant environmental benefits typically "ride the coattails" or occur incidentally as a result of these production-focused efforts. Lean production techniques often create a culture of continuous improvement, employee empowerment, and waste minimization, which is very compatible with organizational characteristics encouraged under environmental management systems (EMS), pollution prevention (P2), and green manufacturing.

Bibliography

Ang, D. S. 1998. Identifications of part families and bottleneck parts in cellular manufacturing. *Industrial Management & Data Systems*, 98(1), 3–7.

Ang, D. S. 2000. An algorithm for handling exceptional elements in cellular manufacturing systems. *Industrial Management & Data Systems*, 100(6), 251–254.

Ang, D. S., and Hegji, C. E. 1997. An algorithm for part families identification in cellular manufacturing. *International Journal of Materials and Product Technology*, 12(4–5), 320–329.

Ang, D. S., McDevitt, C. D., and Jamshidi, H. F. 1994. Machine-part family formation in cellular manufacturing: A computer model. *Business Research Yearbook Global Business Perspectives*, 1, 788–792.

Burbidge, J. L. 1971. Production flow analysis. *Production Engineer*, 50, 139–152.

Burbidge, J. L. 1977. A manual method for production flow analysis. *Production Engineer*, 56, 34–38.

Chan, H. M., and Milner, D. A. 1982. Direct clustering algorithm for group formation in cellular manufacture. *Journal of Manufacturing Systems*, 1, 64–76.

Chandrasekaran, M., and Rajagopalan, R. 1986. MODROC: An extension of rank order clustering for group technology. *International Journal of Production Research*, 24(5), 1221–1223.

Cheng, C. H., Goh, C. H., and Lee, A. 2001. Designing group technology manufacturing systems using heuristic branching rules. *Computers & Industrial Engineering*, 40, 117–131.

Chow, W. S., and Hawaleshka, O. 1992. An efficient algorithm for solving the machine chaining problem in cellular manufacturing. *Computers and Industrial Engineering*, 22(1), 95–100.

Dunlap, G. C., and Hirlenger, C. R. 1983. Well planned coding, classification system offer company-wide synergistic benefits. *Industrial Engineering*, 15(1), 78–83.

ElMaraghy, H. A., and Gu, P. 1988. Knowledge-based system for assignment of parts to machines. *International Journal of Advanced Manufacturing Technology*, 3(1), 33–44.

Güngör, Z., and Arikan, F. 2000. Application of fuzzy decision making in part machine grouping. *International Journal of Production Economics*, 63, 181–193.

Gupta, T., and Seifoddini, H. 1990. Production data based similarity coefficient for machine-component grouping decisions in the design of a cellular manufacturing system. *International Journal of Production Research*, 28, 1247–1269.

Hyer, N. L. 1984. The potential of group technology for U.S. manufacturing. *Journal of Operations Management*, 4(3), 183–202.

Kern, G. M., and Wei, J. C. 1991. The cost of eliminating exceptional elements in group technology cell formation. *International Journal of Production Research*, 29(8), 1535–1547.

King, J. R., and Nakornchai, V. 1982. A machine-component group formation in group technology: A review and extension. *International Journal of Production Research*, 20(2), 117–133.

Kusiak, A., and Chow, W. S. 1987. Efficient solving of the group technology problem. *Journal of Manufacturing Systems*, 6(2), 117–124.

Logendran, R. 1990. A workload based model for minimizing total intercell and intracell moves in cellular manufacturing. *International Journal of Production Research*, 28, 913–925.

Mansouri, S. M., Mooattar, H., and Newman, S. T. 2000. A review of the modern approaches to multi-criteria cell design. *International Journal of Production Research*, 38(5), 1201–1218.

McAuley, J. 1972. Machine grouping for efficient production. *The Production Engineer*, 51, 53–57.

McCormick, E. J., Jenneret, P. R., and Mecham, R. C. 1972. A study of job characteristics and job dimensions as based on the Position Analysis Questionnaire (PAQ). *Journal of Applied Psychology*, 79, 262–268.

Seifoddini, H. 1989. Duplication process in machine cells formation in group technology. *IIE Transactions*, 21(4), 382–388.

Shafer, S., Kern, G., and Wei, J. 1992. A mathematical programming approach for dealing with exceptional elements in cellular manufacturing. *International Journal of Production Research*, 30, 1029–1036.

Singh, N. 1993. Design of cellular manufacturing systems: An invited review. *European Journal of Operational Research*, 69(3), 284–291.

Tsai, C. C., Chu, C. H., and Barta, T. 1997. Analysis and modeling of a manufacturing cell formation problem with fuzzy integer programming. *IIE Transactions*, 29(7), 533–547.

Vannelli, A., and Kumar, K. R. 1986. A method for finding minimal bottle neck cells for grouping part-machine families. *International Journal of Production Research*, 24, 387–400.

Wang, J. X. 2002. *What Every Engineer Should Know about Decision Making under Uncertainty*. Boca Raton, FL: CRC Press.

Wang, J. X. 2005. *Engineering Robust Designs with Six Sigma*. Upper Saddle River, NJ: Prentice Hall.

Wang, J. X. 2010. *Lean Manufacturing: Business Bottom-Line Based*. Boca Raton, FL: CRC Press.

Wang, J. X. 2012. *Green Electronics Manufacturing: Creating Environmental Sensible Products*. Boca Raton, FL: CRC Press.

Wang, J. X., and Roush, M. L. 2000. *What Every Engineer Should Know about Risk Engineering and Management*. Boca Raton, FL: CRC Press.

Wemmerlove, U., and Hyer, N. L. 1986. The part family/machine group identification problem in cellular manufacturing. *Journal of Operation Management*, 6, 125–147.

Won, Y. 2000. Two-phase approach to GT cell formation using efficient p-median formulations. *International Journal of Production Research*, 38(7), 1601–1613.

chapter two

Sequence-based cellular manufacturing

Decision making under risk and uncertainty is applied to cellular manufacturing, specifically in the machine cell formation step. The application works with parts demand, which can be either expressed in a probability distribution (probabilistic production volume) or not expressed in probability distribution, where only the different possible values for volume that can occur are known (uncertain production volume). Decision making under risk is used to help the designer select the best cell arrangement in case of probabilistic production volume. Decision making under uncertainty is used to help the designer select the best cell arrangement in case of uncertain production volume. The objective of the design methodology has been to maximize the profit imposed by the resource capacity constraints.

2.1 Cellular manufacturing: Workplace design based on work cells

As the foundation of cellular manufacturing, the philosophy of group technology (GT) is to identify similar parts and group them together in families to take advantage of their design and manufacturing similarities. This technique is applied to cellular manufacturing where dissimilar machines are grouped together into cells to produce a family of parts. This process is also called machine cell formation. Decision making has been studied from a number of different theoretical approaches. Normative theories focus on how to make the best decisions by deriving algebraic representations of preference from idealized behavioral axioms. Descriptive theories adopt this algebraic representation but incorporate known limitations of human behavior. Computational approaches start from a different set of assumptions altogether, focusing instead on the underlying cognitive and emotional processes that result in the selection of one option over the other.

In this chapter, decision theory under risk and uncertainty is applied to cellular manufacturing specifically in the machine cell formation step. The methodology will address the dynamic nature of the production environment, which can be divided into two situations: a probabilistic production volume and an uncertain production volume. This chapter focuses on manufacturing engineering's decision making under risk and uncertainty. Whether we like it or not, we all feel that the world is uncertain. From choosing a new technology to selecting a manufacturing process, we rarely know in advance what outcome will result from our decisions. Unfortunately, the standard theory of choice under uncertainty developed in the early 1940s and 1950s turns out to be too rigid to take many tricky issues of choice under uncertainty into account. The good news is that we have now moved away from the early descriptively inadequate modeling of behavior. This chapter explains the accomplished progress in individual decision making through the most recent contributions to uncertainty modeling and behavioral decision making. It also introduces the reader to the many subtle issues that need to be resolved for rational choice under uncertainty.

Decision making is certainly the most important task of an engineer and it is often a very difficult one. The domain of decision analysis models falls between two extreme cases. This depends upon the degree of knowledge we have about the outcome of our actions. One "pole" on this scale is deterministic. The opposite "pole" is pure uncertainty. Between these two extremes are problems under risk. The main idea here is that for any given problem, the degree of certainty varies among managers depending upon how much knowledge each one has about the same problem. This reflects the recommendation of a different solution by each person. Probability is an instrument used to measure the likelihood of occurrence for an event. When probability is used to express uncertainty, the deterministic side has a probability of one (or zero), while the other end has a flat (all equally probable) probability. This chapter offers a decision-making procedure for solving complex problems step by step. It presents the decision analysis process for both public and private decision making, using different decision criteria, different types of information, and information of varying quality. It describes the elements in the analysis of decision alternatives and choices, as well as the goals and objectives that guide decision making. The key issues related to a decision-maker's preferences regarding alternatives, criteria for choice and choice modes, together with risk assessment tools, are also presented.

Probabilistic production volume refers to the nature of product volume, which is known and can be expressed in a probabilistic distribution. The uncertain production volume, on the other hand, refers to the nature of production volume, which cannot be expressed in a probabilistic distribution because the distribution of demand is unknown. In this situation, only the different values for volume that can occur are known (i.e., the likelihood of any particular occurrence is unknown).

Decision making under risk and manufacturing is a design methodology. The expected result of this design methodology is a preliminary cellular system design that performs well and more realistically models the real demand situation in manufacturing compared to the constant demand assumption methods currently in use. Making decisions under conditions of risk and uncertainty is a fundamental part of every engineer's and manager's job, whether the situation involves product design, investment choice, regulatory compliance, or human health and safety. This chapter will present both qualitative and quantitative tools for structuring problems, describing uncertainty, assessing risks, and reaching decisions, using a variety of case studies that are not always amenable to standard statistical analysis. Bayesian methods will be introduced, emphasizing the natural connection between probability, utility, and decision making.

2.2 Decision making under risk

When making choices between risky options, human decision makers exhibit a number of framing effects. Decision makers tend to evaluate a gamble relative to a reference point, and seek risk when prospects are framed as losses, but avoid risk when identical prospects are framed as gains. This tendency for risk preferences to reverse between loss and gain frames has been termed the reflection effect and is one of the primary predictions of prospect theory.

2.2.1 Decision matrices for cellular manufacturing

The integration of cost into decision matrices in engineering and engineering technology programs is necessary to provide graduating engineers the skills to become immediate contributors to the goals and profits of their chosen companies. There are many methods that can be used, including sophisticated decision science techniques. One example of a decision science technique is a trade-off based on the strength of the materials. In this example, a simple selection model, called a decision matrix, is used to decide what combination of factors provided a cost-effective design. This type of technique will allow teaching professionals to introduce and reinforce the concept of cost into basic mechanical engineering design courses. Decision matrices are used extensively in many fields of science, health care, electronics, and manufacturing. Obviously, more sophisticated decision science processes using complex analysis and modeling techniques can also be used. The purpose of this chapter is to introduce the concepts of cost and optimization of design to engineering students early in their educational careers. The more sophisticated decision science models can be incorporated into advanced engineering design courses. Techniques such as this can also be applied in many other courses and disciplines including project management. The cost of oil and gasoline has tripled in the last few years. The concept of cost-effective design has suddenly reached new heights and now affects almost everyone in the world. There are four components in a decision matrix (Kleinfeld, 1993):

- Decisions or actions
- States of nature
- Probabilities of the states of nature
- Payoffs

Risk visualization has the potential to reduce the seemingly irrational behavior of decision makers. A model of decision making under risk enhances our understanding of visualization, which can be used to support risk based decision making. Decision making under risk stems from the fact that decision-making scenarios in manufacturing are characterized by uncertainty and a lack of structure. The complexity inherent in such scenarios is manifested in the form of:

- Unavailability of information
- Too many alternatives
- Inability to quantify alternatives
- Lack of knowledge of the payoff matrix

This is particularly prevalent in domains such as investment decision making. Rational decision making in such domains requires a careful assessment of the risk–reward payoff matrix. However, individuals cope with such uncertainty by resorting to a variety of heuristics. Prior decision support models have been unsuccessful in dealing with complexity and nuances that have come to typify such heuristic-based decision making.

2.2.2 *Decisions or actions*

Decisions or actions refer to the array of alternatives that the decision maker identifies. The choice for an aggregation level is determined by the required level of detail of information for the decisions that have to be taken. The choice for an abstraction level can be based on the cost of timely acquisition of the required information versus the cost of omitting part of the information in the analysis.

For example, virtual cellular manufacturing (VCM), an alternative approach to implementing cellular manufacturing, has been investigated. VCM combines the setup efficiency typically obtained by group technology (GT) cellular manufacturing (CM) systems with the routing flexibility of a job shop. Unlike traditional CM systems in which the shop is physically designed as a series of cells, family-based scheduling criteria are used to form logical cells within a shop using a process layout. The result is the formation of temporary, virtual cells as opposed to the more traditional, permanent, physical cells present in GT systems. Virtual cells allow the shop to be more responsive to changes in demand and workload patterns. Production using VCM is compared to production using traditional cellular and job-shop approaches. Results indicate that VCM yields significantly better flow time and due date performance over a wide range of common operating conditions, as well as being more robust to demand variability.

2.2.3 *States of nature*

States of nature refer to all the possible situations that can happen in the future. In other words, these are the major sources of risk in the decision. Flexibility is emerging as one of the key competitive strengths in today's manufacturing systems. Measuring the flexibility of manufacturing systems is important to operations managers engaged in decision making on strategic issues related to flexibility. Flexibility is widely recognized as a multidimensional attribute. Most of the studies reported in the literature have focused on measuring separate dimensions independently and also have tended to be nonfinancial in nature. As a result, many of these measures only have a limited application in strategic decision making. In this paper we consider the measurement of the aggregate flexibility of a manufacturing system by investigating the joint effect of the flexibilities on a variety of dimensions when measurement for a medium-term time horizon is appropriate. We propose a value-based approach in which flexibility is measured by the ability of a manufacturing system to consistently generate high net revenues across all conceivable states of the nature in which it may be called upon to function. We suggest the ratio of the mean to the standard deviation of the distribution of optimal net revenues as the flexibility measure.

2.2.4 *Probabilities of the states of nature*

Probabilities of the states of nature refer to the probability distribution of the states of nature. Of greater importance are two conceptual issues. A matrix such as the one portrayed here is technically referred to as a decision under risk, as most decision theorists define it. Risk means that we have reasonable estimates of the probabilities of the states of nature, and we are therefore engaged in somewhat of a "gambling" problem.

Most real-world decisions have unclear or low-confidence estimates of these probabilities. For these real-world decisions, how can we even be reasonably sure that the probability of sustained demand growth is .70? Why not a 50-50 chance, or a .10 probability if the decision maker feels pessimistic. These conditions, where we cannot estimate the probabilities of the states of nature, are termed decisions under uncertainty by formal decision theorists. Uncertainty is a more difficult problem to resolve than risk, primarily because it permits less direct application of mathematical methods.

2.2.5 *Payoffs*

Payoffs are estimates of the quantitative results associated with each action, conditional upon the outcome of a particular state of nature. Manufacturing decision making is almost always accompanied by conditions of uncertainty. Clearly, the more information the decision maker has, the better the decision will be. Treating decisions as if they were gambles is the basis of decision theory. This means that we have to trade off the value of a certain outcome against its probability. To operate according to the canons of decision theory, we must compute the value of a certain outcome and its probabilities; hence, determining the consequences of our choices. The origin of decision theory is derived from economics by using the utility function of payoffs. It suggests that decisions be made by computing the utility and probability, the ranges of options, and also lays down strategies for good decisions.

Risk implies a degree of uncertainty and an inability to fully control the outcome or consequences of such an action. Risk or the elimination of risk is an effort that engineers employ. However, in some instances the elimination of one risk may increase some other risks. Effective handling of a risk requires its assessment and its subsequent impact on the decision process. The decision process allows the decision maker to evaluate alternative strategies prior to making any decision. The process is as follows:

1. The problem is defined and all feasible alternatives are considered. The possible outcomes for each alternative are evaluated.
2. Outcomes are discussed based on their monetary payoffs or net gain in reference to assets or time.
3. Various uncertainties are quantified in terms of probabilities.
4. The quality of the optimal strategy depends upon the quality of the judgments.

The decision maker should identify and examine the sensitivity of the optimal strategy with respect to the crucial factors.

2.2.6 The principles for decisions under risk

2.2.6.1 The principle of expectation

According to the principle of expectation, the alternative that should be selected is the one that has the minimum expected cost (or maximum expected profit). Thus, an alternative is selected if it has minimum expected cost (or maximum expected profit).

The actual outcome will not equal the expected value. What you get is not what you expect, that is, the "great expectations!"

1. For each action, multiply the probability and payoff.
2. Add up the results by row.
3. Choose the largest number and take that action.

Decisions can be portrayed with a payoff table.

2.2.6.1.1 The payoff table As shown in Table 2.1, the payoff table has a column for each act and a row for each event. The payoff value expresses how closely an outcome brings the decision maker to his or her goal.

2.2.6.1.2 Making the decision using a payoff table Decision makers want to maximize payoff. But, the outcome is uncertain. Various criteria exist for making the choice. Maximizing the expected payoff is commonly used to make choices.

- First, compute the expected payoff for each act. An act's expected payoff is the weighted average value for its column found by multiplying payoffs by the row's probability and summing the products.
- Second, choose the act having the maximum.

Maximizing the expected payoff is called the Bayes decision rule. It is not a perfect criterion because it can lead to the less preferred choice.

Table 2.1 Example of a Payoff Table

		ACT (choice of movement)					
		Gears and Levers		Spring Action		Weights and Pulleys	
EVENT (level of demand)	PROBABILITY	Payoff	Payoff × Probability	Payoff	Payoff × Probability	Payoff	Payoff × Probability
Light	.10	$ 25,000	$ 2,500	–$ 10,000	–$ 1,000	$ 125,000	–$ 12,500
Moderate	.70	400,000	280,000	440,000	308,000	400,000	280,000
Heavy	.20	650,000	130,000	740,000	148,000	750,000	150,000
	Expected Payoff		$412,500		$455,000		$417,500

Table 2.2 Example: Principle of Most Probable Future

EVENT (level of demand)	PROBABILITY	ACT (choice of movement)					
		Gears and Levers		Spring Action		Weights and Pulleys	
		Payoff	Payoff × Probability	Payoff	Payoff × Probability	Payoff	Payoff × Probability
Light	.10	$ 25,000	$ 2,500	–$ 10,000	–$ 1,000	$ 125,000	–$ 12,500
Moderate	.70	400,000	280,000	440,000	308,000	400,000	280,000
Heavy	.20	650,000	130,000	740,000	148,000	750,000	150,000

2.2.6.2 *The principle of most probable future*

Considering only the state that has the highest probability of occurring, we select an action that yields maximum profit or minimum cost.

This method is a simple way for decision making under risk, but it is good for nonrepetitive decisions. The steps of this method are as follows:

1. Take the state of nature with the highest probability (subjectively break any ties). For the example shown in Table 2.2, there is a 70% chance of a "moderate level of demand." Thus, we should take this state of nature.
2. For that state of nature, choose the action with the greatest payoff. For the example shown in Table 2.2, corresponding to "moderate level of demand," Spring Action is the action with the greatest payoff. So we should choose Spring Action with a payoff of $440,000 and an expected payoff of $308,000.

2.2.6.3 *The aspiration-level principle*

The aspiration-level principle requires a decision maker to set a goal or level of aspiration. An alternative is selected if it maximizes the probability that the goal will be met or exceeded.

In most cases, decision makers will set a certain aspiration level before evaluating the alternatives. An interpretation of this philosophy in terms of a decision under risk is to select an alternative that maximizes the probability of achieving the desired aspiration level. For a typical aspiration level, which reflects a decision-maker's trade-off between accomplishing goals and averting risks, there should be (1) a certain internal rate of return (IRR) level, (2) a certain cost level, or (3) a certain profit level.

A useful tool in evaluating a capital project investment is the IRR. This must be calculated by trial and error and is the interest rate that produces a net present value of zero for any period. For a favorable project, the IRR will be higher than the cost of capital. The more conservative, modified internal rate of return, MIRR, is considered an improvement by some because reinvestment at the cost of the capital rate of return is used.

However, net present value (NPV) is usually preferred since it is easier to use, easier to understand, and more conservative than relying on the IRR.

For example, a profitable chemical manufacturer wants to expand its operation at an existing site. Cash is available to invest. Its research and engineering departments are capable of developing the basic data, developing the conceptual and final design, constructing the manufacturing facility, and bringing the new chemical plant on-line successfully. So enough engineering has been completed to estimate the project cost, plant management is able to estimate the mill cost, and enough market development has been completed to forecast sales revenue for a number of years. The plant is expected to show a good return on investment, ROI, which is calculated by dividing the earnings by the total investment.

2.3 *Decision making under uncertainty*

In the previous section, we considered decision problems in which the decision maker does not know the consequences of his choices but is given the probability of each consequence under each choice. In most economic applications, such a probability is not given. For example, for manufacturing planning in a competitive business setting, an engineer cares not only about what her or his company plans but also about what the other companies (competitors) plan. Hence, the description of consequences includes the strategy profiles. In that case, in order to fit in that framework, we would need to give other competitors' mixed strategy profiles in the description of the manufacture planning, making simulation theoretical analysis moot.

In all these problems, the decision makers hold subjective beliefs about the unknown aspects of the problem and use these beliefs to make their decisions. For example, an engineer proposes her or his strategy according to her or his beliefs about what other companies (competitors) may plan, and the engineer may reach these beliefs through a combination of reasoning and the knowledge of past behavior. This is called decision making under uncertainty. The components of a decision matrix for decision making under uncertainty are the same as those for decision making under risk except that there are no probabilities of the states of nature. Manufacturing decision making under uncertainty can be characterized as follows:

- The decision maker has to choose one act from a given set of possible acts.
- A set of potential states is given, representing the circumstances about which there is uncertainty.
- It is logically necessary that one and only one of the states will occur; otherwise, the states have been erroneously specified.
- Eventually, the uncertainty will be resolved and one state will be realized.

2.3.1 *The minimax principle*

Minimax is a decision rule used in decision theory, game theory, statistics, and philosophy for minimizing the possible loss for a worst-case (maximum loss) scenario. Alternatively, it can be thought of as maximizing the minimum gain (maximin). This principle is based on the view that the worst possible outcome will occur. The alternative that minimizes the maximum cost or maximizes the minimum profit is selected. Minimax can be summarized as follows:

- A principle for decision making by which, when presented with two various and conflicting strategies, one should, by the use of logic, determine and use the strategy that will minimize the maximum losses that could occur.
- This business strategy strives to attain results that will cause the least amount of regret, should the strategy fail.

In summary, in situations with conflicting alternatives, the most rational strategy is the one that promises to minimize the maximum possible losses.

2.3.2 *The minimin (or maximax) principle*

The minimin (or maximax) principle is based on the view that the best possible outcome will occur. The alternative that minimizes the minimum cost or maximizes the maximum profit is selected. The background of the minimin (or maximax) principle can be summarized as follows:

- Pontryagin's maximum (or minimum) principle is used in optimal control theory to find the best possible control for taking a dynamical system from one state to another, especially in the presence of constraints for the state or input controls.
- It was formulated in 1956 by the Russian mathematician Lev Semenovich Pontryagin and his students.
- It has as a special case the Euler–Lagrange equation of the calculus of variations.
- The minimum principle provides necessary conditions, but not sufficient conditions, for optimality.

2.3.3 *The Hurwicz principle*

The minimax or maximin principle is extremely pessimistic. The minimin or maximax principle is extremely optimistic. The Hurwicz principle allows selection between these two extremes. Based on the Hurwicz principle, engineers should select the alternative that has the largest weighted average of its maximum and minimum payoffs.

2.3.4 The Laplace principle

The Laplace principle is based on the view that all future outcomes are equally likely to occur. The alternative that yields the minimum expected cost or maximum expected profit is selected. The Laplace principle suggests choosing a strategy that is optimal in a situation where the opponent chooses all strategies with equal probabilities. In other words, according to the Laplace principle, the best that we can do under uncertainty is to behave as under risk, where all strategies of the opponent might appear with equal probabilities.

2.3.5 The Savage principle (minimax regret)

The Savage principle is based on the view that a decision maker is interested in the difference between the actual outcome and the outcome that could have happened. The alternative that minimizes the maximum difference is selected. The Savage principle can be summarized as follows:

- It is a technique used in decision theory.
- A criterion is used to construct a regret matrix in which each outcome entry represents a regret defined as the difference between the best possible outcome and the given outcome.
- The matrix is then used as in decision making under risk with the expected regret as the decision-determining quality.
- It is also known as the regret criterion.

2.4 Sequence-based materials flow procedure

Sequence-based materials flow procedures have been developed to consider operation sequences in accurately determining the costs of intercell movement, as well as forward and backward intracell movements.

One of the most important problems in designing a cellular manufacturing system is the formation of machine cells and the grouping of parts. Most of the existing cell design procedures operate on a binary part–machine incidence matrix to identify block diagonality. Important factors such as part operation sequence and demand volume have not been commonly considered in past research. In this section, a sequence-based materials flow procedure is developed to solve the cell formation problem. The new cell formation procedure is designed to consider operation sequences in accurately determining the costs of intercell movement, as well as forward and backward intracell movements. Extensive comparisons of the new cell formation procedure with the relevant existing procedures in the literature demonstrate the effectiveness of the new procedure.

First, each machine is considered as a cell. Then the merge of a pair of cells that yields the most negative change in total cost is made. The total materials flow cost includes the cost of intercell movement, forward intracell movement, backtracking, and skipping. The procedure iterates until the cost reduction by merging two cells cannot be made.

2.4.1 Notations of describing sequence-based materials flow

The notations of the sequence-based materials flow procedure are as follows:

n_m, n_p, n_c	Number of machines, parts, and cells, respectively
N_{ij}	Operation number of part i on machine j ($N_{ij} = 0$, if part i does not visit machine j)
$\propto_i$	Intercell material handling cost for part i ($/unit)
β_i	Intracell material handling cost for part i ($/unit)
γ_i	Backtracking the cost multiplier
θ	Machine skipping cost ($/unit/skip)
q_i	Number of units of part i that need to be processed
SC_i	Total skipping cost in cell i
H	Change in forward flow intracell cost
M	Change in intercell cost
S	Change in skipping cost
B	Change in backward flow intracell (backtracking) cost
r_{ek}	Net increase in total materials flow cost if cell e is combined with cell k to form a new cell

2.4.2 Algorithm for assessing sequence-based materials flow

The steps of the sequence-based materials flow procedure are as follows:

1. $n_c = n_m$ (initially each machine is considered as one cell). Let $e = 1$.

2a. Let $k = e + 1$.

2b. Find the total skipping costs in the combined cell ek, and individual cells e and k. In each cell, count the number of skips for part 1 to n_p. Then calculate the change in overall skipping cost, $S = Sc_{ek} - (Sc_e + Sc_k)$.

2c. Let B = H = M = 0.
Consider all possible pairs of machines of which one machine belongs to cell e and the other belongs to cell k. According to the forward flow direction of the pair of machines, let the earlier machine index be $m1$ and the later machine index be $m2$, that is, $f_{m1} < f_{m2}$.
For $i = 1$ to n_p
  While $N_{im1} > 0$ and $N_{im2} > 0$, if $(N_{im1} - N_{im2}) = 1$
  $B = B + q_i\gamma_i\beta_i$ and $M = M - q_i\alpha_i$, if $(N_{im1} - N_{im2}) = -1$
  $H = H + q_i\beta_i$ and $M = M - q_i\alpha_i$

Consider the next pair of machines. If all pairs have been considered, go to step 2d.

2d. $r_{ek} = S + H + M + B$
$k = k + 1$; if $k > n_c$, go to step 3, or else go to step 2b.

3. $e = e + 1$; if $e = n_c$, go to step 4, or else go to step 2a.
4. Find the pair of cells, x and y, that correspond to the most negative r_{ek} value, that is, r_{xy} = Min $[r_{ek}]$; if $r_{xy} > 0$, stopping criterion is reached, go to step 6, or else go to step 5.
5. Combine cells x and y.
$n_c = n_c - 1$, $e = 1$, if $n_c = 1$, go to step 6, or else go to step 2a to start the next iteration.
6. Allocate each part of the cell in which it has the maximum number of operations.
7. Stop.

An operation sequence-based method integrates intracell layout cost with cell formation to minimize the total cost of the materials flow and machine investment. The sequence-based method has been developed for designing a cell.

2.5 Problem formation

Cell formation, one of the most important stages in CM, is to group parts with similar geometry, function, material, and process into part families and the corresponding machines into machine cells. The design of cellular manufacturing systems is a complex, multicriteria, and multistep process. In the design of cellular manufacturing systems, design objective(s) must be specified. Minimizing intercell moves, distances, costs, and the number of exceptional parts (parts that need more than one cell for processing) are common design objectives. An exceptional part can be also called an exceptional element or a bottleneck part. The problem is defined in terms of the design objective, system parameters, and system constraints. The following information is assumed available.

1. The set of machines, M = $\{M_1, \ldots, Mn_m\}$ and their capacities, CM_j, $j = 1, \ldots, n_m$.
2. The set of all parts, $P = \{P_1, \ldots, Pn_p\}$.
3. Parts demand with different values for part volume that can occur are known.
4. The processing sequence and processing time of corresponding machine for each part.
5. The profit for producing each part.
6. The transportation cost (\$/unit) (both intracell and intercell) and skipping cost (\$/unit/skip) for each part.

2.5.1 The design objective

Each design approach considers different numbers of design objectives and constraints to a different extent, depending upon the scope and interest of each. For instance, clustering analysis approaches consider only one objective, the minimization of intercell moves. In the design process of clustering techniques, only part operations and the machines for processing those operations are considered. Other product data (such as operational sequences and processing times) and production requirements (such as production rate) are not incorporated into the design process. Thus, solutions obtained may be valid in limited situations. However, they are simple to implement and solutions can be obtained in reasonable amounts of time. Minimizing the intercell flow of parts is fundamental to achieving many of the benefits associated with CM. The cell formation problem is complicated by the existence of exceptional parts or exceptional machines. Both exceptional parts and exceptional machines cause the intercellular movement of parts. Ideally, a part-cluster is processed in a single machine cell for its entire operation. In practice, however, it is a very rare case. Extensive work has been done by many researchers to provide new techniques for solving this problem. The design objective is to develop a new machine cell formation methodology in such a way that the profit is maximized.

2.5.2 System parameters

2.5.2.1 Part demand

Part demand is defined as the quantity of each part in the product mix to be produced. In this chapter, different values for part volume that can happen for each part are known.

2.5.2.2 Operation sequence

The operation sequence is defined as an ordered list of the machine types that the part must visit to be operated. Considering the operation sequence provides a more realistic and accurate baseline to determine the cost of intercell movement, as well as forward and backward intracell movement.

2.5.2.3 Processing time

Processing time is defined as the time required for operating a part. Normally, setup and run time are included in processing time. The processing time should be provided for every part on corresponding machines in the operation sequence. Processing time is used to determine resource capacity requirements.

2.5.2.4 Resource capacity

The resource capacity is defined as the machine time available for producing parts. When dealing with many possible demands, we need to consider whether the resource capacity is violated.

2.5.2.5 Material handling cost

Material handling cost is defined as the cost per unit for moving a part between machines. There are two kinds of movement involved in material handling cost.

- Intercell movement cost is the cost incurred when a part moves from one cell to another cell.
- Intracell movement cost which is subdivided into the cost incurred when a part moves within a cell.
- Forward flow cost is the cost incurred when a part moves from one machine to another machine inside a cell in the forward direction.
- Backward flow cost is the cost incurred when a part moves from one machine to another machine inside a cell in the backward direction.

The direction of the flow can be found by defining an average weighted operation number, f_k, for each machine k. f_k of machine k is defined by

$$f_k = (\Sigma_i N_{ik} q_i)/(\Sigma q_i)$$

where

N_{ik} is the operation number of part i on machine k

q_i is the demand volume of part i

The forward flow direction can be found by the order of the machines in ascending order of the f_k values.

The material handling cost can be calculated by

$$\sum_{i=1}^{np} \left(H_i q_i \beta_i + B_i q_i \gamma_i + M_i q_i \alpha_i \right)$$

where

i = part number

q_i = number of units of part i

$H_i q_i \beta_i$ = forward intracell material handling cost for part i

H_i = number of forward intracell movement for part i

β_i = intracell material handling cost for part i ($/unit)

$B_i q_i \gamma_i$ = backtracking cost for part i

B_i = number of backtracks for part *i*
γ_i = backtracking cost multiplier
$M_i q_i \alpha_i$ = intercell material handling cost for part *i*
M_i = number of intercell movements for part *i*
α_i = intercell material handling cost for part *i* ($/unit)

2.5.2.6 Skipping cost

Skipping cost is the cost incurred when some parts do not visit each machine in the cell, resulting in special handling and increased delays, and may require larger work-in-process (WIP). Skipping cost can be calculated by

$$\sum_{i=1}^{np}\sum_{m=1}^{nc}(SC_{im})\theta q_i$$

where

i = part number
m = cell number
Sc_{im} = number of machine skips for part *i* in cell *m*
θ = skipping cost ($/unit/skip)
q_i = number of units of part i

2.5.2.7 Total material flow cost

Total material flow cost is defined as the cost comprising material handling cost (intercell movement, forward intracell movement, and backtracking) and skipping cost.

Total material flow cost = Material handling cost + Skipping cost

2.5.2.8 Total profit

Total profit can be defined as

Total profit = Profit (without material flow cost) –
Total material flow cost

2.5.3 System constraints

The constraint in this chapter is the resource capacity or machine time available. The feasible part volume vector (q_1, ..., q_{np}) needs to be under resource capacity, or machine time available need to be determined such that the overall profit is maximized. The optimal production volume is determined by solving the following linear programming problem.

$$\text{Maximize } \beta_1 q_1 + \ldots + \beta_{np} q_{np}$$

subject to

$$\sum_{i=1}^{np} q_i \Theta_{ij} \le CM_j, j = 1, \ldots, nm$$

$$0 \le q_i \le d_i, i = 1, \ldots, np$$

where

The objective function ($\beta_1 q_1 + \ldots + \beta_{np} q_{np}$) is the total profit.
β_i, $i = 1, \ldots, np$ is the unit profit of producing part i.
q_i, $i = 1, \ldots, np$ is the feasible volume for part i.
Θ_{ij} = the processing time of part i on machine j.
d_i, $i = 1, \ldots, np$ is the demand for part i.

In today's economic climate, many organizations struggle with declining sales and increasing costs. Some choose to hunker down and weather the storm, hoping for better results in the future. However, layoffs and workforce reductions jeopardize future competitiveness. While organizations that have implemented the theory of constraints (TOC) continue to thrive and grow in difficult times, continuing to achieve real bottomline growth, whether by improving productivity or increased revenues. The TOC adopts the common idiom "a chain is no stronger than its weakest link" as a new management paradigm. This means that processes, organizations, and so forth, are vulnerable because the weakest person or part can always damage or break them or at least adversely affect the outcome. The TOC applies the cause-and-effect thinking processes used in the hard sciences to understand and improve all systems, but particularly organizations.

2.6 *Machine cell formation under probabilistic demand*

The data for machine cell formation in cellular manufacturing is organized in a machine–component chart that represents the machining requirements of parts in the product mix. While the composition of the product mix is determined by demand and is probabilistic in nature, the existing machine cell formation models treat it as deterministic. This could adversely affect the performance of the associated cellular manufacturing system. To overcome this problem, probabilistic machine cell formation models have been developed.

1. Calculate the joint probability of the part volumes.
2. For every part volume combination, determine the feasible part volumes, $q_1, \ldots, q_{np}$, which maximize the profit from the linear programming problem mentioned in Section 2.5.3.
3. For every feasible part volume, one machine cell arrangement is designed. In this chapter, a sequence-based materials flow procedure developed by Verma and Ding (1995) is used.
4. For every machine cell arrangement:
   a. Calculate the total material flow cost including the cost components of intercell movement, forward intracell movement, backtracking, and skipping for every feasible part volume.
   b. Calculate the total profit, which can be defined as

Total profit = Profit [from step 2] – Total material flow cost [from step 4a]

5. Build a profit decision matrix.
   - *States of nature* are all feasible part volume combinations found in step 2.
   - *Probabilities of the states of nature* are the joint probabilities found from step 1.
   - *Action or decisions* are all cell arrangements found from step 3. Payoffs can be calculated by subtracting the total material flow cost (step 4a) from the profit (step 2).
   - *Payoffs* are the profit of all feasible part volume correspondents to each cell arrangement found from step 4b.
6. Apply the principle for decisions under risk to select the cell arrangement.

2.7 Machine cell formation under risk

Whenever the decision maker has some knowledge regarding the states of nature, he or she may be able to assign subjective probability for the occurrence of each state of nature. By doing so, the problem is then classified as decision making under risk. In many cases, the decision maker may need an expert's judgment to sharpen his or her uncertainties with respect to the likelihood of each state of nature. In such a case, the decision maker may buy the expert's relevant knowledge in order to make a better decision. The procedure used to incorporate the expert's advice with the decision maker's probabilities assessment is known as the Bayesian approach.

A classical decision problem is considered where a decision maker is to choose one of a number of actions each offering different consequences. The outcome from a choice of action is uncertain because it depends on the existing state of nature. Also, the outcome, once an action and state of nature are specified, may be a vector or a random vector. The decision maker employs both Bayesian methods and fuzzy set techniques to handle the uncertainties. The decision maker is also allowed to use multiple, possibly conflicting, goals to determine the best strategy. The Bayesian method produces a set of undominated strategies to choose from, whereas the fuzzy set technique usually produces a unique optimal strategy.

2.8 Machine cell formation under uncertainty

Problem formulation and assumptions of the machine cell formation under uncertainty are the same as those of the machine cell formation under probabilistic demand. However, the distribution of demand is unknown and as a result joint probability of part volume combination cannot be found.

When the probability distributions of part demand are unknown, decision making under uncertainty can be used. The algorithm is similar to that of the cell formation under risk except the principle for selecting the cell arrangement used is the principle for a decision under uncertainty. NP stands for nondeterministic polynomial time. An NP-complete problem cannot be solved in polynomial time in any known way. The machine–part cell formation is an NP-complete combinational optimization in a cellular manufacturing system. Previous research has revealed that although the genetic algorithm (GA) can obtain high quality solutions, special selection strategy, crossover, and mutation operators as well as the parameters must be previously defined to solve the problem efficiently and flexibly. The estimation of distribution algorithms (EDAs) has recently been recognized as a new computing paradigm in evolutionary computation that can overcome some drawbacks of the traditional GA mentioned earlier.

2.9 Six Sigma

In this chapter, decision theory applied to cellular manufacturing in machine cell formation step is presented. The application works with part demand, which can be either expressed in a probability distribution or not expressed in probability distribution, where only the different possible values for volume that can occur are known. The objective of design methodology has been to maximize the profit imposed by the resource capacity constraints.

Estimation of the failure probability, that is, the probability of unacceptable system performance, is an important and computationally challenging problem in manufacturing reliability engineering. In cases of practical interest, the failure probability is given by an integral over a high-dimensional uncertain parameter space. Over the past decade, the engineering research community has realized the importance of advanced stochastic simulation methods for solving manufacturing reliability problems.

Six Sigma consists of a set of statistical methods for systemically analyzing processes to reduce process variation, which are sometimes used to support and guide organizational continual improvement activities. Six Sigma's toolbox of statistical process control and analytical techniques are being used by some companies to assess process quality and waste areas to which other Lean methods can be applied as solutions. Six Sigma is also being used to further drive productivity and quality improvements in Lean operations.

Six Sigma was developed by Motorola in the 1990s, drawing on well-established statistical quality control techniques and data analysis methods. The term *sigma* is a Greek alphabet letter (σ) used to describe variability. A sigma quality level serves as an indicator of how often defects are likely to occur in processes, parts, or products. A Six Sigma quality level equates to approximately 3.4 defects per million opportunities, representing high quality and minimal process variability.

It is important to note that not all companies using Six Sigma methods are implementing Lean manufacturing systems or using other Lean methods. Six Sigma has evolved among some companies to include methods for implementing and maintaining performance of process improvements. The statistical tools of the Six Sigma system are designed to help an organization correctly diagnose the root causes of performance gaps and variability, and apply the most appropriate tools and solutions in addressing those gaps.

Bibliography

Askin, R. G., and Standridge, C. R. 1993. *Modeling and Analysis of Manufacturing Systems*. New York: John Wiley & Sons, Inc.

Buchanan, J. T. 1982. *Discrete and Dynamic Decision Analysis*. New York: John Wiley & Sons.

Buck, J. R. 1989. *Economic Risk Decisions in Engineering and Management*. Ames, IA: Iowa State University Press.

Fleischer, G. A. 1994. *Introduction to Engineering Economy*. Boston: PWS Publishing.

Harhalakis, G., Ioannou, G., Minis, I., and Nagi, R. 1994. Manufacturing cell formation under random product demand. *International Journal of Production Research*, 32(1), 47–64.

Kleinfeld, I. H. 1993. *Engineering Economics Analysis for Evaluation of Alternatives*. New York: Van Nostrand Reinhold.

Kusiak, A. 1990. *Intelligent Manufacturing Systems*. Englewood Cliffs, NJ: Prentice Hall.

Longendran, R. 1991. Impact of sequence of operations and layout of cells in cellular manufacturing. *International Journal of Production Research*, 29(2), 375–390.

Nagi, R., Harhalakis, G., and Proth, J.-M. 1990. Multiple routings and capacity considerations in group technology applications. *International Journal of Production Research*, 28(12), 1243–1257.

Offodile, O. F., Mehrez, A., and Grznar, J. 1994. Cellular manufacturing: A taxonomic review framework. *Journal of Manufacturing Systems*, 13(3), 196–220.

Okogbaa, O. G., Chen, M.-T., Changchit, C., and Shell, R. L. 1992. Manufacturing system cell formation and evaluation using a new inter-cell flow reduction heuristic. *International Journal of Production Research*, 30(5), 1101–1118.

Seifoddini, H. 1990. A probabilistic model for machine cell formation. *Journal of Manufacturing Systems*, 9(1), 69–75.

Vakharia, A. J., and Wemmeröv, U. 1990. Designing a cellular manufacturing system: A materials flow approach based on operation sequences. *IIE Transactions*, 22(1), 84–97.

Verma, P., and Ding, F. Y. 1995. A sequence-based materials flow procedure for designing manufacturing cells. *International Journal of Production Research*, 33(12), 3267–3281.

Wang, J. X. 2002. *What Every Engineer Should Know about Decision Making under Uncertainty*. Boca Raton, FL: CRC Press.

Wang, J. X. 2005. *Engineering Robust Designs with Six Sigma*. Upper Saddle River, NJ: Prentice Hall.

Wang, J. X. 2010. *Lean Manufacturing: Business Bottom-Line Based*. Boca Raton, FL: CRC Press.

Wang, J. X. 2012. *Green Electronics Manufacturing: Creating Environmental Sensible Products*. Boca Raton, FL: CRC Press.

Wang, J. X., and Roush, M. L. 2002. *What Every Engineer Should Know about Risk Engineering and Management*. Boca Raton, FL: CRC Press.

Wicks, E. M. 1995. Designing a cellular manufacturing system with time varying product mix and resource availability. Doctoral Dissertation, Virginia Polytechnic Institute and State University.

chapter three

Cellular manufacturing and jidoka (autonomation)

In today's business world, competitiveness defines an industry leader. The drive toward maximum efficiency is constantly at the forefront of such companies' objectives. By incorporating jidoka (autonomation), engineers are striving to adopt Lean manufacturing practices to help address worries about their bottom line. Cellular manufacturing is one staple of Lean manufacturing.

3.1 Overall equipment effectiveness for cellular manufacturing

A manufacturing system is a collection of integrated equipment and human resources, whose function is to perform one or more processing or assembly operations on a starting raw material, part, or set of parts. Cellular manufacturing is an approach that helps build a variety of products with as little waste as possible. A cell is a group of workstations, machine tools, or equipment arranged to create a smooth flow so families of parts can be processed progressively from one workstation to another without waiting for a batch to be completed or requiring additional handling between operations. Put simply, cellular manufacturing groups together machinery and a small team of staff, directed by a team leader, so all the work on a product or part can be accomplished in the same cell eliminating resources that do not add value to the product.

The manufacturing system is where value-added work is performed to parts or products, and this activity gives manufacturing a central place in the overall scheme of the system of production, where it is supported by systems of manufacturing support, quality control, material handling, and automation control.

Different types of manufacturing systems may be identified. For cellular manufacturing, effective management of OEE (overall equipment effectiveness) is the key to the prevention of costly downtime at a high value plant. Sensors are available to provide valuable information in the quest to improve OEE, but to date there are few comprehensive systems that allow plant owners to readily measure OEE and take prompt and proactive action to improve it. Together with proper maintenance and inspection regimes, OEE has been shown to improve equipment availability and greatly reduce unplanned outages. In order to manage OEE, large amounts of data need to be collected from a wide range of sources, which then needs to be processed. This can be time consuming and potentially prone to error unless great care is taken.

Manufacturing systems consist of human workers, automation, and various material handling technologies, configured in ways that create specific manufacturing system typologies. More specifically, a manufacturing system is a collection of integrated equipment and human resources, whose function is to perform one or more processing or assembly operations on a starting raw material, part, or set of parts. Our focus in this chapter is on manufacturing systems that are automated, and so the concentration will be on the types of integrated equipment that are used and arranged in a manufacturing cell. This can range from production machines and tools, material handling, and work positioning devices to the use of various computer systems that facilitate automation in the production environment.

For cellular manufacturing, a machine cell is a series of manually operated production machines and workstations, often in a U-shaped configuration, that performs a sequence of operations on a family of parts or products that are similar but not identical. A flexible manufacturing system (FMS) is a highly automated machine cell that produces part or product families; it often consists of workstations comprising computer numerical control (CNC) machine tools.

3.2 *Jidoka (autonomation) for ensuring overall equipment effectiveness*

The most common definition of *jidoka* is "autonomation." The word is one of many Japanese terms that are common in Lean manufacturing. In cellular manufacturing, production workstations and equipment are arranged in a sequence that supports a smooth flow of materials and components through the production process with minimal transport or delay. Implementation of this Lean method often represents the first major shift in production activity, and it is the key enabler of increased production velocity and flexibility, as well as the reduction of capital requirements. Lean jidoka traces its roots to the early 1900s at Toyota in Japan, then a textile manufacturing company. Sakichi Toyoda developed a device that could detect broken threads in a loom and stop the machine from producing defective material. The concept of jidoka enabled companies to greatly increase the number of machines a single operator could run, with very little extra effort on the worker's part. With jidoka, Lean becomes much easier for operators and much more profitable for companies.

The jidoka definition mentioned earlier (autonomation) is essentially automation with a human touch. But it has also come to mean more than that. It is about stopping production whenever an abnormal condition is detected, fixing the defect, and then using countermeasures to prevent further occurrences. Many jidoka devices are combined with an andon light, or signaling device, to alert the operator of the abnormal condition. The purpose of jidoka is to separate people from machines, so an operator can do more while the machine is running. Rather than processing multiple parts before sending them on to the next machine or process step (as is the case in batch-and-queue, or large-lot production), cellular manufacturing aims to move products through the manufacturing process one piece at a time, at a rate determined by the customers' needs. Cellular manufacturing can also provide companies with the flexibility to vary product type or features on the production line in response to specific customer demands. This approach seeks to minimize the time it takes for a single product to flow through the entire production process.

Jidoka is often one of the pillars of a company's production system (the Lean house). Just-in-time (JIT; the other common pillar) and jidoka work together to create manufacturing excellence. This one-piece flow method includes specific analytical techniques for assessing current operations and designing a new cell-based manufacturing layout that will shorten cycle and changeover times. To make the cellular design work, an organization must often replace large, high-volume production machines with small, flexible, "right-sized" machines to fit well in the cell. Equipment often must be modified to stop and signal when a cycle is complete or when problems occur.

This transformation often shifts worker responsibilities from watching a single machine to managing multiple machines in a production cell. Although plant floor workers may need to feed or unload pieces at the beginning or end of the process sequence, they are generally free to focus on implementing total productive maintenance (TPM) and process improvements. By using this technique, production capacity can be incrementally increased or decreased by adding or removing production cells.

3.3 *Supervisory control and data acquisition (SCADA) for implementing jidoka*

Supervisory control and data acquisition (SCADA) systems are widely used in the utilities and manufacturing industries to monitor and control production and distribution of products. OEE management can benefit from the application of SCADA techniques. SCADA solutions could be used to cost-effectively automate the collection of plant data and compute key performance indicators (such as availability) in real time to provide the information required to allow manufacturing organizations or equipment manufacturers to manage OEE.

SCADA systems are utilized to assist in the operation of all energy infrastructures (electric power, oil, and natural gas), as well as in water, waste water, and factory floor automation systems. SCADA provides the ability to monitor operational parameters at remote facilities and to adjust physical controls at those facilities in a timely manner, in most cases without the need for on-site personnel.

Through the use of automation, the energy infrastructure utilizes SCADA systems to operate key infrastructure components such as

- Generating stations
- The electric grid
- Petroleum production platforms
- Pipeline systems
- Petroleum refineries
- Compressor stations
- Storage facilities

This chapter provides an introduction to SCADA, including a brief discussion of its history and evolution. It also discusses the importance of SCADA to energy infrastructure operations and examines SCADA vulnerabilities that exist.

3.3.1 SCADA for energy efficiency management

Older plants that have been repeatedly added to or modified often suffer from inefficient material flow. In addition, efficient material flow is more critical in plants with continuous processes than batch processes. The implementation process of shedding the traditional manufacturing processes and embracing the drastically different cellular manufacturing techniques can be a daunting task. Management must deal with many issues including cell design and set up, team design and placement, employee training, and teamwork training, as well as other company functional issues. A project team should be put together that consists of management and production employees to handle these changes. SCADA has been adopted by energy infrastructure operators as a tool to assist them in efficiently and cost-effectively managing their complex individual systems. This is done through measuring, monitoring, and controlling key functions at nodes within the infrastructure. The number of individual nodes that have operational significance within a single energy transmission or distribution system may be in the thousands, and may be widely scattered geographically. Without automation, management of a system of this size would require the efforts of a large work force, both in the field and in the office, and would of necessity mandate a considerable degree of decentralized operational control. SCADA assists in the operation and management by providing a computer-based technology that offers flexibility in operational efforts, ease in gathering and maintaining data, and the ability to develop studies and reports on a timely basis.

Before there was SCADA there was telemetry. Telemetry is a technique developed years earlier by the railroad industry to monitor locations of rolling stock, as well as the status of switches on the track system. Using electric wires, lights, and relays, telemetry allowed a central dispatcher to efficiently and safely schedule traffic on the rail network. Initially these communications were transmitted over the telegraph system, which was the earliest wired communications system in commercial use.

The energy infrastructures (oil, gas, and electricity) were comparable to railroads in that they had complex remote facilities, large capital investments that required protection, and relatively limited and simple control requirements. Telemetry was adaptable to their needs and offered a means of reducing operating costs. As a result, telemetry was in wide use in all branches of the energy industry by the early 1960s. These early systems communicated with the remote facilities primarily through the same telegraph or telephone lines used for day-to-day communications. At the same time, radiotelemetry was developing, and some systems began using this new technology, especially in remote areas, where it was difficult and expensive to install physical facilities.

Telemetered data was processed by relays installed at the central control facility. As systems grew, the relay systems became more complex, numbering into the thousands. Space and maintenance requirements became limiting factors in designing telemetry systems. Digital computers were introduced to data gathering in the early 1960s. This new technology gave system designers the freedom to centralize data gathering for very large systems. When midsize units (so-called minicomputers) became available in about 1965, they provided the first capability for true two-way communications between the remote field units and the central control facility. This was the beginning of SCADA, although the term was not yet in use.

The term SCADA was first used in the early 1970s, and the word telemetry was used less frequently. SCADA described a system that included a computerized control facility that could both receive and send data along a communications path between itself and numerous remote field devices. At about the same time, advancements in radio technology made wireless communications a reliable and cost-effective alternative for two-way systems. Most new SCADA systems began using radio communications.

The development of the minicomputer, along with progress in software and telecommunications capability, initially provided a means to observe system operations, anticipate changes in demand, and make appropriate adjustments without the need for on-site personnel. The consequent reduction in manpower and other operating costs led the energy infrastructures to adopt and rely on SCADA systems. Today, these systems have become quite sophisticated and powerful as computer capabilities have advanced.

SCADA is a computer system for gathering and analyzing real-time data. SCADA systems manage parts inventories for cellular manufacturing, regulate industrial automation and robots, and monitor process and quality control. SCADA has been used to monitor and control utility networks and manufacturing processes and provide asset performance information to allow business decisions to be made. Standard solutions now exist that are field proven to a very large scale, and these deliver high reliability and repeatability in a cost-effective manner.

3.3.2 Components of a SCADA system

Cell design and setup should be executed to facilitate the movement of the product through its production cycle and should also be able to produce other similar products as well. The cells are arranged in a manner that minimizes material movement and are generally set up in a U-shaped configuration.

A SCADA system in its simplest form consists of a central host or master computer (usually called a master station, master terminal unit or MTU) communicating with one or more data gathering units or remotes (usually called remote stations, remote terminal units, or RTUs) by means of one or more communications media and a collection of standard or custom software. Most systems provide a level of redundancy by utilizing multiple computers and partial or total duplication of communications paths to maintain constant contact between the control center and each of the field units.

A wide range of telecommunications media (i.e., landlines, public switched networks, radio frequencies, VHF/UHF, microwaves, cellular telephones, fiber optic cable, and satellites) are utilized to link RTUs and the MTU together. In many systems, the communications function is provided by a commercial communications entity over a public communications system (e.g., phone, cell phone, VSAT). In others, the operator owns and maintains the communications as a dedicated part of the SCADA system (e.g., VHF/UHF radio, microwave). In addition to these necessary components, SCADA systems also incorporate certain corollary devices to facilitate distribution, use, and maintenance of the data that is obtained. Examples of these include databases, graphic displays, and off-line processing capability.

Figure 3.1 provides a diagram of the interconnectivity of a SCADA system. The control facility is where the master computer, or MTU, is

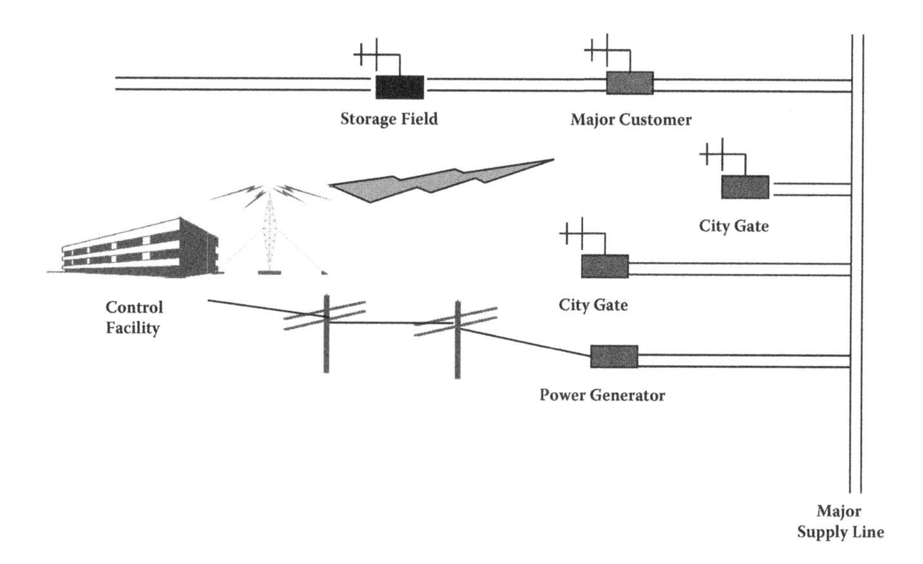

Figure 3.1 SCADA integrates control of remote facilities.

located. The field devices (RTUs) can be found at critical components such as power generators, city gate stations, pipelines, compressor and pumping stations, storage facilities, and major end-use customers.

3.3.2.1 The master terminal unit

The master terminal unit (MTU) is at the heart of any SCADA system, or more precisely, it is the brain. The level of computing capability in the MTU determines the limit of sophistication possible for any planned automation activities. The MTU is operated through a man–machine interface (M/MI), which is a means of enabling the operator to issue commands through the MTU to the facilities in the field. This almost always includes a keyboard, but may also include mouse-driven graphic displays, touch screens, and the like.

Most SCADA systems are designed with one (or more) computer units functioning as the primary MTU, with at least one additional unit installed in parallel and programmed to act as a backup. The standby unit typically is up and continuously running, and receives all incoming data at the same time as the primary unit. It is also used by the control staff for various off-line activities, such as compiling and issuing reports, archive retrieval, and what-if studies. In most cases, it is identical to the primary unit, although it merely needs to have approximately equal capabilities. In the event the primary unit fails the standby unit is available for immediate service. Some systems are designed for the standby unit to automatically take over system control if the primary unit fails. Others require manual intervention and a controlled introduction of the standby unit.

3.3.2.2 Remote standby

Many SCADA operators desire greater security than is provided merely by a backup MTU within the control center. It is common practice among transmission companies, and even some local distribution utilities, to maintain an emergency control center at a point some distance from the main center. The emergency center is equipped with another MTU, as well as whatever peripheral equipment is necessary to allow remote operation for a period up to several days or weeks. These facilities provide continuity of operation in the face of hurricanes, fire, bomb threats, or any event that makes operation from the main control center imprudent or impossible.

Some companies maintain these facilities on a continuously live basis, while others bring them up and online only when in actual use, for test or for real operation. Usually, this decision is predicated on whether the facility is normally manned anyway, such as when the remote site is a compressor station or crew reporting center.

3.3.2.3 A remote terminal unit

A remote terminal unit (RTU) is an instrument that monitors one or more physical parameters at a specific point in a piping system or wire network. Its function is similar to that of a gauge or meter, but it is also capable of passing on, via a modem, the data it is observing. An RTU can be either intelligent or not. The simplest RTU compares the condition it sees (pressure, voltage, temperature, etc.) to a preestablished range and converts it to a digital value that represents the appropriate percentage of that range. The digital value is transmitted when the MTU polls the RTU at regular intervals. Another type of RTU, called an accumulator, counts repetitive signals from a field device and transmits the number of pulses seen since it was last polled by the MTU. These RTUs do nothing but gather data and report it. They cannot institute any status change.

Certain RTUs are designed to receive a signal from the MTU and substitute it for an existing signal at a particular device, such as a valve operator. This causes the device to adjust to a new setting, with a corresponding change in system conditions. None of the RTUs described so far require intelligence to perform its function.

An intelligent RTU is sometimes also referred to as a programmable logic controller (PLC) or an intelligent electronic device (IED). Such a device contains one or more microchips that can be programmed to enable the RTU to perform certain computational or control tasks in the field, relieving the MTU of some of its duties. An example of such an intelligent unit might be a flow computer as used in most gas transmission SCADA systems.

Simple (nonintelligent) RTUs can collect raw data regarding gas conditions across an orifice plate in a meter tube, and forward the data to the

MTU. The MTU then processes the data using standard flow equations contained in its programming and calculates the instantaneous flow. It may be required to do this for dozens of points on the system, each time the system is scanned. This requires significant computing time.

An intelligent field unit gathers the appropriate raw data and makes the flow calculation on-site, using the same equations used by the MTU. It then transmits the actual flow value to the MTU. The flow computer can perform this measurement and calculation on a nearly continuous basis, while the MTU can only do it as often as the RTU is polled. The flow computer obtains a more accurate volume for the measurement period, which is then transmitted to the MTU once each day. It also provides a greater measure of security in the event of a communications or MTU disruption, because it continues to store usable data for later transmission. Only the instantaneous raw values are lost. Most SCADA systems employ both simple and intelligent field devices.

3.3.2.4 Communications

In order for a SCADA system to function there must be reliable two-way communications between the MTU in the control center and the RTU in the field. This can be provided by the public switched phone system, fiber optics, cellular telephone, VHF or UHF radio, microwave, satellite radio, or even a dedicated hardwired system. Each has advantages and drawbacks. An operator must consider the impact of these factors relative to his particular system when deciding which communication mode should be used.

Leased space on the public network is reliable and requires little in the way of purchased front-end equipment. If it is imperative that the installed cost of a system be kept to a minimum, this may be the best choice. The savings in initial investment capital will be offset by a recurring expense. There will be a regular monthly bill for each field location, which may include both local and distance charges. These charges can be substantial for a system with many data points. The same is true for a fiber optic system. In addition, both of these alternatives are susceptible to service disruptions due to inadvertent dig-ins by outside forces.

Cellular and satellite systems require more investment up front, and they still are subject to regular bills for service. They are not vulnerable to dig-ins, although cell towers are exposed to possible attack. Microwave and radio systems using lower frequencies are usually owned by the SCADA operator, and do not carry a monthly financial burden. However, each has its own tower infrastructure requirement, representing both an up-front investment and ongoing maintenance costs. Licensing requirements can also impose costs, as well as extend the time required for a new system to become operational. In many cases, professional legal services are necessary for the operator to avoid serious licensing difficulties.

A privately owned hardwired system is practical for only the most localized SCADA systems. They are used in many plant operations, where the system is confined to a single operation. The cost of real estate and construction make it impractical for most utility operations.

It is important to note that there are numerous systems in the communications arena that now use the Internet as the primary communications medium. This is an attractive concept because it is easy to access, can be very cost-effective for far-flung systems (every transmission can be a local call), and is reliable if the operator is judicious in selecting the ISP. However, such a system increases the vulnerability to intrusion. This will be discussed further in Section 4.4 (Chapter 4) on threats and vulnerabilities.

3.3.2.5 *Software*

Today's SCADA systems require sophisticated software to facilitate their operations. These programs provide the necessary operational flexibility, as well as the ability to archive data, create and utilize graphic schematics, and issue regular reports. Some systems even allow off-line simulations for use in operator training programs.

In the early days of SCADA, limited computer power imposed restrictions on the amount of data that could be gathered and processed. Because there were no intelligent field devices, data gathered in the field had to be processed after it was received by the MTU. Most SCADA operating programs were custom programs, developed either in-house or by a software contractor that tailored the program specifically to fit an operator's system and set of requirements. There were therefore many different and incompatible operating protocols and philosophies in use, frequently dictated by the limits of the computer platform on which they ran.

Today, most SCADA systems use commercially available software packages that employ open architecture protocols, making each system look very much like many others. This feature provides ease of installation and setup of a new system. It also facilitates compatibility between devices offered by different vendors. Operators are not limited to single sources for subsequent additions to the system.

Commercial software is designed to utilize the power and versatility available on today's computers, particularly the PC. Interactive graphics, touch screens, and Intranet and Internet connections are all normal parts of today's SCADA systems, and SCADA systems are integrated into the overall information system of the corporation. This expanded accessibility has a concomitant impact on security considerations. This will also be discussed further in Section 4.4 on threats and vulnerabilities in the next chapter.

For cellular manufacturing, team design and placement is a crucial part of the process. Employees must work together in cell teams and are led by a team leader. This team leader becomes a source of support for the cell and is oftentimes responsible for the overall quality of the product that leaves his or her cell.

Employee training must also accompany the change to cellular manufacturing. In cellular manufacturing, workers generally operate more than one machine within a cell, which requires additional training for each employee, which in turn creates a more highly skilled workforce. This cross-training allows one employee to become proficient with his or her machines, while also creating the ability to operate other machines within the cell when such needs arise.

Teamwork training should generate camaraderie within each cell and stimulate group-related troubleshooting. Employees within each team are empowered to employ ideas or processes that would allow continuous improvement within the cell, thus reducing lead times, removing waste, and improving the overall quality of the product.

Other issues that must be addressed include changes in purchasing, production planning and control, and cost accounting practices. Arranging people and equipment into cells help companies meet two goals of Lean manufacturing: one-piece flow and high-variety production. These concepts dramatically change the amount of inventories needed over a certain period of time.

- One-piece flow is driven by the needs of the customer and exists when products move through a process one unit at a time thus eliminating batch processing. The goals of one-piece flow are to produce one unit at a time continuously without unplanned interruptions and without lengthy queue times.
- High-variety production is also driven by the needs of customers who expect customization as well as specific quantities delivered at specific times. Cellular manufacturing provides companies the flexibility to give customers the variety they demand by grouping similar products into families that can be processed within the same cell and in the same sequence. This eliminates the need to produce products in large lots by significantly shortening the time required for changeover between products.

3.4 Case study: Time-based cellular manufacturing

Thermoelectric cooling uses the Peltier effect to create a heat flux between the junction of two different types of materials. A Peltier cooler is a solid-state active heat pump that transfers heat from one side of the device to the other, with consumption of electrical energy, depending on the direction of the current. The growth of Thermal Electric Corporation (TEC) has influenced the purchase of a new manufacturing and warehouse facility.

The layout of the facility must be designed in order to address the issues of implementing Lean manufacturing, inventory reduction, reduction in material handling, and improve efficiencies in the shipping and delivery process.

The main advantages of a Peltier cooler (compared to a vapor-compression refrigerator) are its lack of moving parts or circulating liquid, and its small size and flexible shape (form factor). Its main disadvantage is that it cannot simultaneously have low cost and high power efficiency. The purpose of the TEC case study is to design a warehouse/manufacturing facility. TEC is a manufacturer and distributor of professional audio systems. TEC has currently outgrown its existing facility and contracted the design of a new facility to an engineering consulting company. The new building is 120,000 square feet and will contain offices, manufacturing, distribution and warehousing of finished goods, raw materials, and packaging supplies. TEC anticipates a 7% volume growth per year.

Peltier (thermoelectric) cooler performance is a function of ambient temperature, hot and cold side heat exchanger (heat sink) performance, thermal load, Peltier module (thermopile) geometry, and Peltier electrical parameters. The goal of the design team is to propose a manufacturing, assembly, and distribution center using proper material handling, inventory, and operational practices for the next 5 years. A current problem for TEC is that it carries excessive inventory. The design team will reduce all inventories to a cost-effective more realistic quantity. The team also will create an effective cellular manufacturing process for the speakers to aid in the direction of Lean manufacturing. Finally, the team will address the issue of shipping for TEC.

3.4.1 Problem statement

Thermoelectric coolers are commonly used in camping and portable coolers, and for cooling electronic components and small instruments. Some electronic equipment intended for military use in the field is thermoelectrically cooled. The cooling effect of Peltier heat pumps can also be used to extract water from the air in dehumidifiers. TEC has grown out of its current location and has purchased a new facility to keep up with demand; the problem is that it has no direction as far as the layout of this new facility. Its present production configuration is inefficient and does not utilize the ideas of Lean manufacturing. Problems with the grouping of similar types along with the flow of production need to be addressed in order to optimize the process. There is also a problem with the method in which the finished products are being distributed. Currently, a third-party trucking company is in charge of delivery; this system is also very inefficient with all the trucks leaving with less than full loads. It also has an abundance of finished goods pallets; this inventory does not make good use of the floor space available and it costs money.

3.4.2 Plan for cellular manufacturing

A camping/car type electric cooler can typically reduce the temperature by up to 20°C below the ambient temperature. With feedback circuitry, thermoelectric coolers can be used to implement highly stable temperature controllers that keep the desired temperature within ±0.01°C. Such stability may be used in precise laser applications to avoid laser wavelength drifting as the environment temperature changes. The plan for TEC is to reduce the inventory over a 5-year period to 20%. This 20% inventory buffer will allow TEC to cover its peak sales times without worrying about running out of inventory. Every aspect of the facility is designed to be a Lean system. This is achieved by minimizing the distance between operations, placing all materials that a cell needs in close proximity to the cell, and strategically planning what operations are performed at each individual cell to maximize efficiency.

When approaching the problem it was a clear decision to use a cellular manufacturing system. A cellular manufacturing system was chosen because all the products assembled are thermoelectric coolers. Having a product all of one family allows workers to increase productivity because only the dimensions of our product change while the end result stays the same. Having a product all in one family also allows the workers to use the same equipment on each thermoelectric cooler model, which reduces setup and throughput time. This in turn allows multifunctional workers within a cell to influence the quality of the product. A cellular manufacturing system also reduces the cost of quality because the workers are responsible for their own work within a cell. Thermoelectric cooling elements are a common component in thermal cyclers, used for the synthesis of DNA by a polymerase chain reaction (PCR), a common molecular biological technique that requires the rapid heating and cooling of the reaction mixture for denaturation, primer annealing, and enzymatic synthesis cycles.

3.4.3 Develop cells

Thermoelectric coolers can be used to cool computer components to keep temperatures within design limits or to maintain stable functioning when overclocking. However, due to low efficiency, much more heat is generated than normal, necessitating a very large and noisy fan or a liquid cooling system. A Peltier cooler with a heat sink or waterblock can cool a chip to well below ambient temperature. For cellular manufacturing, the designing of cells has a large impact on product times and product quality. These cells were designed to have a powered conveyor to feed each individual cell of the product. The tables the workers are working on are hydraulic scissor lifts that also have a 30-degree hydraulic tilt lift so workers can access the entire front and back of the product without reaching over the tables. The tops of the hydraulic tables are gravity rollers so the workers can glide the product onto the powered conveyor to go to the next workstation instead of pushing it across a flat surface. All of the materials needed at each workstation are behind the workers separated into bins to increase efficiency.

The most efficient plan is to have two lines of four individual cells and an end cell where the products are tested, packaged, and labeled for finished goods storage. This is arranged by first sorting the dimensions of each speaker produced and then placing them into a large speaker and small speaker category. These two categories will be produced on separate assembly lines to reduce setup time when changing jobs and increase overall efficiency. The four individual cells are then broken down by the time it takes to perform each task. Cell 1 consists of wood assembly only. This is because after the assembly of the wood it takes 60 minutes for the glue to dry. After the speakers are assembled in cell 1 they are placed onto a 30-inch wide power conveyor that is 60 feet long and travels at 1 foot per minute to give them the required 60 minutes of drying time before reaching cell 2. Cell 2 is the carpet attachment area. This is a single-process cell because of the long assembly time it takes for the carpet attachment. Cell 3 places the speaker into the box and wires the speaker box. Cell 4 closes the back of the speaker and attaches all the accessories. Cell 5 has two feeds from the two different cell 4 assembly lines going to a single cell 5. A single fifth cell was chosen because of the short time it takes to test each speaker and then carton, label, and palletize them. After a full pallet has been assembled, a forklift takes the pallet away to be wrapped in cellophane for stability and then places it where the ASC system determines the assembly's identification via his RFID scanner. Note however, that after year 6, a second fifth cell will need to be added as the company increases production.

3.4.4 *Alternative manufacturing systems*

An alternative system to cellular manufacturing is to run either a flow-shop or job-shop type environment. A flow shop would have been undesirable because even though each worker would only have one or two responsibilities, the setup time would increase when changing over cooler sizes, the workers could become more vulnerable to repetitive motion injuries, and production was high enough for each type of thermoelectric cooler. TEC does not produce enough speakers per day for this type of operation to be efficient and cost-effective. A job-shop manufacturing system was not chosen because there is very little variability of the product and the product needed to have an organized way of completing each type of thermoelectric cooler before starting the next type.

3.4.5 Establish inventory control

Because TEC has such a large starting inventory, a feasible solution had to be created for placing the most frequently used products in an easily accessible manner to optimize order picking time. A warehouse management system (WMS) called ASC was a complete package that solved most of TEC's problems for $50,000. The benefits of the ASC system are as follows:

Receiving Module
- Compares expected receipts with received goods to report discrepancies
- Directed put away to tell operators where to put received goods

Picking Module
- Optimized picking for efficiency and accuracy
- Directed picking to tell the operator what to pick and the location from which to pick it
- Directed picking of both raw materials and finished goods
- Scheduled customer orders, pickers, and dock doors
- Support for batch and wave picking, as well as many other picking variations/combinations
- Generates bills of lading, compliance shipping labels, and shipping reports
- Supports kitting and critical cross-docking

Inventory Management Module
- Zone control, replenishment, moves, repacks, and more
- Provides lot traceability, code date, and expired date logic for FIFO (first in, first out)
- Directs operators to conduct cycle and physical counts as part of regular daily work
- Features a no-shutdown "floating physical" count

Because this warehouse management system is so powerful, it organizes all of the products into an ABC inventory control style for TEC. The system then cycles through the oldest products first so the company does not run into a last in, first out situation. The entire system runs on RFID technology. RFID tags eliminate the need for barcode scanning in the warehouse and essentially make it possible to locate every piece of inventory in the entire warehouse from raw materials to finished goods. RFID technology also makes it possible to keep real-time inventory control for raw materials so that an order can be placed up to the minute when a specific raw material has reached its reorder point. The ASC system with RFID technology also makes it extremely easy for order picking. When filling an order, a forklift driver has a handheld RFID scanner with a built-in computer screen that displays all of the information for the upcoming job including where all of the thermoelectric coolers are located in the warehouse and the most efficient way to retrieve the speakers for the current job. This will decrease order picking time because all of the routes for the forklift driver will be optimized by the computer.

3.4.6 Identify alternative inventory control methods

One alternative inventory control method that was not chosen because of cost was a pick-to-light system. A pick-to-light system is still arranged in an ABC style of inventory control but instead of RFID technology telling the driver where to go, it is done by a lighting system on the pallet racks. On a particular order, the driver punches the order number into his handheld scanner, and a light then illuminates where the first speakers are in the order. When the forklift driver is finished picking up that line in the order, the next line for the order lights up, and so on. This type of system was not implemented because pick-to-light systems are designed for warehouses that have extremely large and diverse products and the systems usually cost more than $1 million to integrate and install.

3.4.7 The demand

The production demand considered in this study is adapted from Wicks and Reasor (1999). The plan horizon for which the demand is forecast extends to 3 years with three 1-year periods. A total of 25 different types of parts are to be manufactured, with the product mix and demand volume changing from period to period. Each part type requires processing on different machines in a definite sequence. These machine sequences and the corresponding average processing times in minutes for all the part types are given in Table 3.1. It also provides the average demand for all part types in each period.

3.4.8 Develop plant layout

In the process layout, machines of the same type form a department. In each department, the number of servers needed to meet the processing and setup time requirements of all parts are provided and are obtained as

Table 3.1 Part Data

Part type number	Machine sequence	Processing time (min.)	Annual demand		
			Period 1	Period 2	Period 3
1	10-1-9	7-35-14	300	200	500
2	5-8	42-28	700	600	500
3	1-2-11	7-21-28	0	600	400
4	3-10-6	7-7-42	0	700	800
5	2-5-9	21-7-28	800	600	1000
6	5-10-8	28-35-42	600	300	0
7	6-5-10	21-42-14	0	900	800
8	4-9-11	42-42-7	400	800	200
9	6-10-11	14-42-21	300	200	600
10	3-11	14-28	400	1000	500
11	3-1-4	42-21-28	0	0	200
12	7-9	21-7	700	700	1000
13	3-1-5	42-28-14	100	600	800
14	7-8-10	7-21-21	100	200	0
15	3-9-4	21-14- 7	0	0	300
16	4-10	21-42	500	800	500
17	6-5	42-21	100	900	400
18	1-6-10	14-21-21	1000	1000	400
19	3-6-5	28-7-21	0	700	1000
20	11-9-4	21-28-42	800	300	500
21	8-7	7-14	500	400	1000
22	10-2-11	35-21-14	0	100	100
23	9-6-10	7-42-21	400	500	800
24	7-2	14-28	0	0	500
25	2-7-6	35-42-14	0	0	400

shown in Table 3.2. The layout of departments is modeled as a quadratic assignment problem and is solved using a simulated annealing algorithm. The layout is shown in Figure 3.2.

3.4.9 *The adaptive cellular layout system*

For the adaptive cellular layout system, the best configuration for part families and machine cells for different periods is the one that minimizes the total cost for material handling, machine acquisition, and machine relocation. It is modeled as an optimization problem with an objective function that minimizes the sum of intercellular material handling costs, machine acquisition costs, and machine relocation costs for the plan horizon. It is

Table 3.2 Servers in a Process Layout

Machine type	1	2	3	4	5	6	7	8	9	10	11
Number of servers	1	1	1	1	2	2	1	1	1	2	1

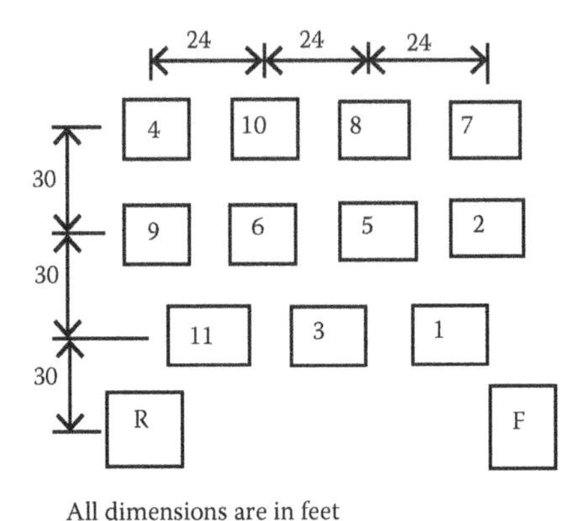

Figure 3.2 Process layouts.

solved using a genetic algorithm (GA) with the approaches described in Wicks and Reasor (1999) and Pillai and Subbarao (2008). The number of cells is taken as three.

In each cell, the required number of servers to meet the processing and setup time requirements of parts are provided. The layout of different cells and also that of machines within each cell in the cellular layout is modeled as a quadratic assignment problem and is solved using a simulated annealing algorithm. Table 3.3 shows the cell and part family configuration and Figures 3.3, 3.4, and 3.5 show the layouts obtained for the three periods.

3.4.10 *The robust cellular layout system*

For the robust cellular layout system the best configuration for part families and machine cells is modeled as an optimization problem. The total cost to be minimized only consists of the intercellular material handling cost and machine acquisition cost. Solution procedures used for the configuration and layout are the same as those for the adaptive design. Table 3.4 shows the cell and part family configuration, and Figure 3.6 shows the layout.

Table 3.3 Cells and Part Families in an Adaptive Layout

	Cell 1	Cell 2	Cell 3
		Period 1	
Machine type (number of servers)	5(1), 7(1), 8(1)	2(1), 5(1), 6(1), 8(1), 9(1), 10(1), 11(1)	1(1), 3(1), 4(1), 5(1), 6(1), 7(1), 9(1), 10(1), 11(1)
Part families	2, 14, 21	5, 6, 9, 17, 23	1, 8, 10, 12, 13, 16, 18, 20
		Period 2	
Machine type (number of servers)	5(1), 7(1), 8(1)	2(1), 5(1), 6(1), 8(1), 9(1), 10(1), 11(1)	1(1), 2(1), 3(1), 4(1), 5(1), 6(1), 7(1), 9(1), 10(1), 11(1)
Part families	2, 14, 21	5, 6, 7, 9, 17, 22, 23	1, 3, 4, 8, 10, 12, 13, 16, 18, 19, 20
		Period 3	
Machine type (number of servers)	5(1), 7(1), 8(1), 9(1)	5(1), 6(1), 10(1), 11(1)	1(1), 2(1), 3(1), 4(1), 5(1), 6(1), 7(1), 9(1), 10(1), 11(1)
Part families	2, 12, 21	7, 9, 17	1, 3, 4, 5, 8, 10, 11, 13, 15, 16, 18, 19, 20, 22, 23, 24, 25

3.4.11 Simulation methodology

The performance of the manufacturing systems in undertaking the given demand is studied using discrete event simulation. A simulation model that is flexible to accommodate a wide variety of manufacturing systems is built. The simulation program is coded using MATLAB® programming features.

3.4.11.1 The simulation model

The simulation model is built in a modular manner. The part data, such as the operation sequence, the machine and average processing time required for each operation, and the demand for each part, are given as input to the simulation. The part family grouping and the machine cell composition are also given as input. The information about layout in the form of a distance matrix showing rectilinear distances between the centers of different machines, the raw material, and finished goods stores, is also given as input.

New batches for production are released at the beginning of every week. The batch size is assumed to be Poisson distributed with the mean being equal to the average weekly demand for the part. The material

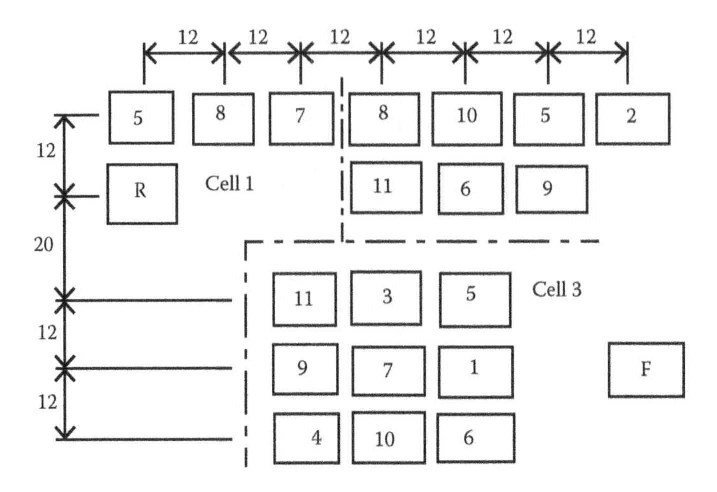

Figure 3.3 Adaptive layout for the first period.

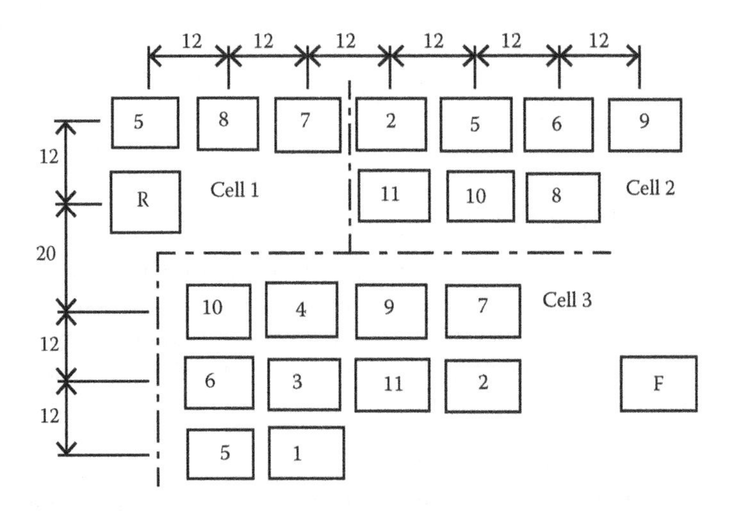

Figure 3.4 Adaptive layout for the second period.

handling system transfers the inventory with a speed of 60 feet per minute, the distance being the rectilinear distance between machines. To account for the time for communicating and waiting for the material handling device, a constant time of 15 minutes is added for intercellular and interdepartmental transfers. There is a common queue for servers of the same type in a cell and in a department of process layout. Parts are taken for processing on a first-come-first-serve basis. The processing time for the operations are assumed to follow k-Erlang distribution with k equal to 2 and mean equal to the average processing time for the operation.

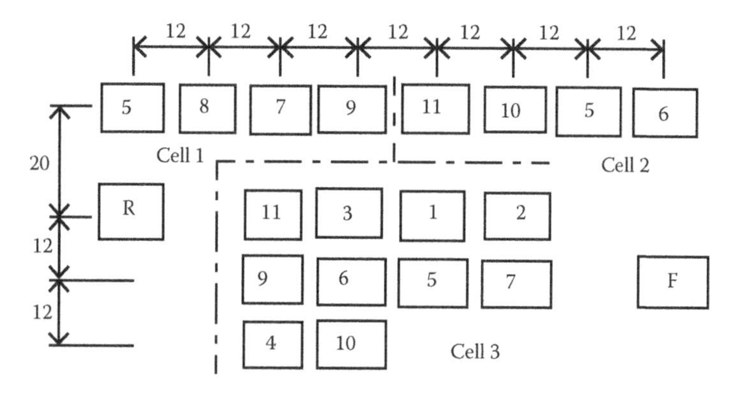

Figure 3.5 Adaptive layout for the third period.

Table 3.4 Cells and Part Families in the Robust Layout

	Cell 1	Cell 2	Cell 3
Machine type (number of servers)	5(1), 6(1), 10(1)	4(1), 9(1), 11(1)	1(1), 2(1), 3(1), 5(1), 6(1), 7(1), 8(1), 9(1), 10(1), 11(1)
Part families in period 1	6, 16, 17	8, 20	1, 2, 5, 9, 10, 12, 13, 14, 18, 21, 23
Part families in period 2	7, 16, 17	8, 20	1, 2, 3, 4, 5, 6, 9, 10, 12, 13, 14, 18, 19, 21, 22, 23
Part families in period 3	7, 16, 17	8, 15, 20	1, 2, 3, 4, 5, 9, 10, 11, 12, 13, 18, 19, 21, 22, 23, 24, 25

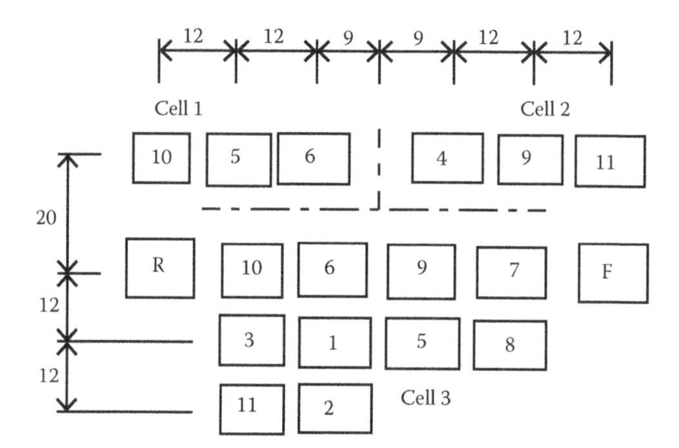

Figure 3.6 Robust layout.

Distributions with a rightward skew are more suitable for the time to complete a task and k-Erlang with $k = 2$ has less variability than exponential distribution (Assad et al., 2003). The setup time is 50% of the processing time of the average batch size. In cellular manufacturing systems, a reduction in setup time is achieved since the cells process similar parts (Wemmerlov and Hyer, 1989). In robust and adaptive layouts, a setup time reduction of 40% is assumed. The manufacturing systems work 8 hours per day, 6 days a week, and 52 weeks per period. In the adaptive system, the cell compositions, machine locations, and part-family allocation all are subject to change at the end of a period. It is difficult to manage a part that is not finished after these changes. They have to be expedited to completion before the period ends. Therefore in this model, a part is not continued for processing after the end of a period in the adaptive layout, but it is continued in the process and robust cellular layouts.

Sufficient replications for obtaining the estimates within ±5% of the mean with 95% confidence level, calculated as 10, are made. The arrival batch sizes for corresponding replications are the same for both layouts.

3.4.11.2 Performance measures and output analysis

The performance measures used here to study the manufacturing systems are production rate, average manufacturing lead time (MLT), and average work-in-process (WIP) inventory. Paired t-tests are conducted to compare the manufacturing systems.

3.4.12 Results

The production rate in the different periods as well as the average production rate for the whole plan horizon, achieved in the three layouts, are given in Table 3.5.

In the adaptive cellular layout, the machines are relocated at the beginning of every period. Therefore in a new period, it will be difficult to complete the processing of incomplete parts of the previous period. The incomplete parts have to be expedited to completion before the machines are relocated. The number of unfinished parts at the end of each period is

Table 3.5 Comparison of Production Rate

	Production rate (parts/day)			
	Period 1	Period 2	Period 3	Overall
Process	24.79	38.92	42.67	35.46
Adaptive	24.81	38.94	42.63	35.46
Robust	24.81	38.94	42.64	35.46

Table 3.6 Parts Unfinished at the End of a Period

	Example 1		
	Period 1	Period 2	Period 3
Process	7.0	16.2	22.5
Adaptive	0.0	2.8	18.5
Robust	0.0	2.1	18.4

given in Table 3.6. For the adaptive layout, these numbers give a measure of the expedition work required.

The average MLT for the three layouts is compared in Figure 3.7, and the comparison of the average WIP inventory is shown in Figure 3.8. Paired *t*-tests are conducted to know if the cellular systems have a performance advantage over the process layout. The level of significance, α,

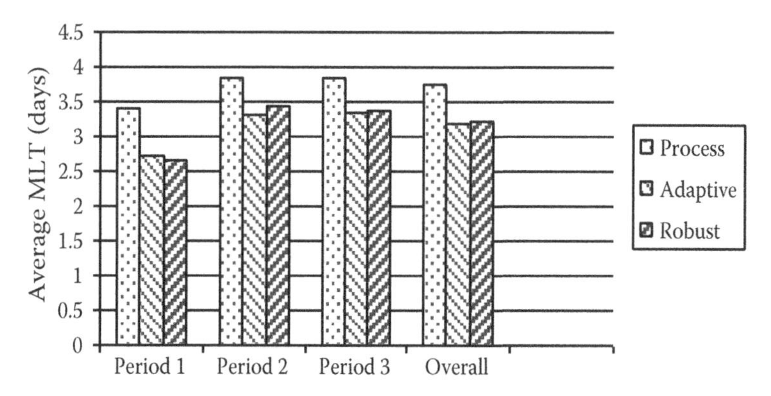

Figure 3.7 Comparison of manufacturing lead time (MLT).

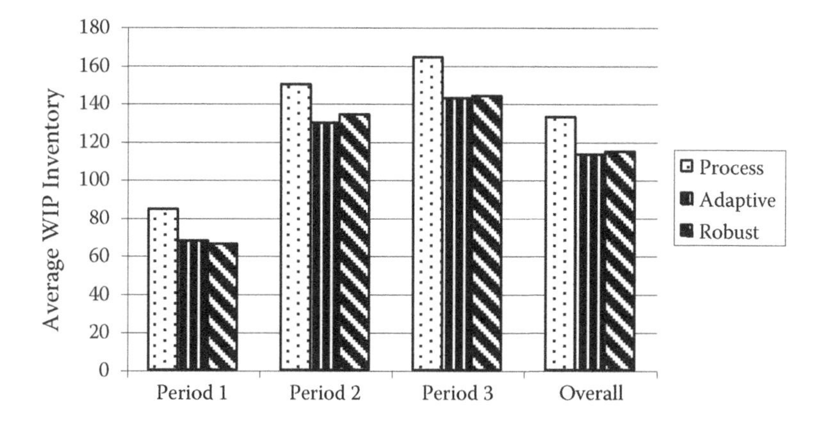

Figure 3.8 Comparison of WIP inventory.

Table 3.7 Details of *t*-Test (Degrees of Freedom = 18, α = 0.05)

	Production/day			MLT (days)			WIP inventory		
	PL	AL	RL	PL	AL	RL	PL	AL	RL
Average	35.460	35.462	35.465	3.741	3.191	3.230	132.79	113.29	114.65
Standard deviation	0.187	0.187	0.188	0.019	0.017	0.022	1.243	0.980	1.157
t_α		1.734	1.734		1.734	1.734		1.734	1.734
T		0.015	0.052		68.19	55.27		38.96	33.77

Notes: PL, process layout; AL, adaptive layout; RL, robust layout.

used is 0.05. The average values (averaged over the plan horizon) of the performance measures, their standard deviations, calculated value of *t*, and the t_α are given in Table 3.7. The differences in production rates are seen to be insignificant. The test shows that the MLT of the adaptive layout is significantly lower than that of the process layout. The MLT of the robust layout is also significantly lower than that of the process layout. The overall MLT for the plan horizon shows a decrease of 14.7% in the adaptive layout and a decrease of 13.7% in the robust layout when compared to the process layout.

The average WIP inventory is lower for the adaptive layout compared to the process layout at a 0.05 significance level. It is also lower for the robust layout compared to the process layout. The reduction of the average WIP inventory level is 14.7% in the adaptive layout when compared to the process layout. For the robust layout, this reduction compared to the process layout is 13.6%.

3.5 *Kaizen, or rapid improvement processes*

Kaizen, or rapid improvement processes, is often considered to be the building block of all Lean production methods. Kaizen focuses on eliminating waste, improving productivity, and achieving sustained continual improvement in targeted activities and processes of an organization.

Lean production is founded on the idea of Kaizen—or continual improvement. This philosophy implies that small, incremental changes applied routinely and sustained over a long period of time result in significant improvements. The Kaizen strategy aims to involve workers from multiple functions and levels in the organization in working together to address a problem or improve a process. The team uses analytical techniques, such as value stream mapping and the 5 Whys, to identify opportunities to quickly eliminate waste in a targeted process or production area. The team works to implement chosen improvements rapidly (often within 72 hours of initiating the Kaizen event), typically focusing on solutions that do not involve large capital outlays.

Periodic follow-up events aim to ensure that the improvements from the Kaizen "blitz" are sustained over time. Kaizen can be used as an analytical method for implementing several other Lean methods, including conversions to cellular manufacturing and just-in-time production systems.

Cellular manufacturing creates the ability to incorporate one-piece flow production, which produces multiple time and monetary benefits. First, it reduces material handling and transit times. By having the machinery to complete a certain process grouped together in a cell, the product spends more time on the machinery and less time in transit between machines. Unlike batch processing, materials do not accumulate at a certain location to be worked or moved. This allows the operator the ability (in most cases) to move the unfinished product to the next station without the need of specialized equipment to move what would be, in a batch process, a larger load farther distances.

Previous studies have shown that the performance of cellular manufacturing systems deteriorates when the demand becomes unstable. But these studies were based on cellular systems designed for stable demand. In this study the performances of two cellular systems that are suitable for multiperiod demands, namely, the adaptive cellular manufacturing system and the robust cellular manufacturing system, and that of a process layout, are analyzed for the same multiperiod demand situation. In cellular systems, a setup time reduction of 40% is used. The cellular layouts are found to perform better than the process layout in performance measures such as average manufacturing lead time and average WIP inventory. Setup reduction and reduced material transfer distances are the reasons for this. It has been shown that cellular systems consistently maintain the performance advantage in all of the periods.

With decreased material handling and transit time, accompanied by virtually eliminating queue times associated with batch processing, comes shortened part-cycle times. In other words, the time to produce one unit of a particular product results in shorter delivery dates for the customer. The effort of relocation of machines, loss of working time due to relocation, and the effort of handling unfinished parts during relocation, which happens in adaptive layout, is avoided in the robust layout. Hence, the robust layout offers a better solution in multiperiod demands.

One-piece flow is also associated with reduced WIP inventories. With a continuous and balanced flow of product through the cell, no major buildup of material occurs between workstations, eliminating the need of excess space to store in-process goods. This also allows workstations and machinery to be moved closer together. Less WIP is easier to manage and allows the manufacturer to operate with shorter lead times. Performance of these cellular systems under various levels of setup reduction has to be analyzed. Such a study could indicate the cutoff level of setup reduction to be achieved for the cellular systems to become a preferable alternative.

Another benefit of cellular manufacturing is based on the capability to produce families of similar products within each cell. Adjustments required to set up machinery should not be significant for each family product. Reduced change over time will enable more frequent product line changes and items can be produced and delivered in smaller lot sizes without significant cost implications.

In addition to the aforementioned production benefits, there are also numerous benefits that are associated with the employees and their involvement in each cell. First, a cell on average employs a small number of workers that produce the complete part or product. Workers become multifunctional and are responsible for operating and maintaining numerous pieces of equipment and workstations. They are also able to cover other workstations within the cell when required to do so.

In terms of worker productivity, the ability to deal with a product from start to finish creates a sense of responsibility and an increased feeling of teamwork. A common purpose is created and gives "ownership" to the production teams. Feedback on quality and efficiency is also generated from the teams building continuous improvement within the cells and adjusting quality issues immediately and not after an entire batch has been produced.

Bibliography

Ang, C. L., and Willey, P. C. T. 1984. A comparative study of the performance of pure and hybrid group technology manufacturing systems using computer simulation techniques. *International Journal of Production Research*, 22(2), 193–233.

Assad, A. A., Kramer, S. B., and Kaku, B. K. 2003. Comparing functional and cellular layouts: A simulation study based on standardization. *International Journal of Production Research*, 41(8), 1639–1663.

Flynn, B. B., and Jacobs, F. R. 1986. A simulation comparison of group technology with traditional job shop manufacturing. *International Journal of Production Research*, 24(5), 1171–1192.

Morris, J. S., and Tersine, R. J. 1990. A simulation analysis of factors influencing the attractiveness of group technology cellular layouts. *Management Science*, 36(12), 1567–1578.

Needy, K. L., Billo, R. E., and Warner, R. C. 1998. A cost model for the evaluation of alternative cellular manufacturing configurations. *Computers and Industrial Engineering*, 34(1), 119–134.

Pillai, V. M., and Subbarao, K. 2008. Robust cellular manufacturing system design for dynamic part population using a genetic algorithm. *International Journal of Production Research*, 46(18), 5191–5210.

Seifoddini, H., and Djassemi, M. 1997. Determination of flexibility range for cellular manufacturing systems under product mix variations. *International Journal of Production Research,* 35(2), 3349–3366.
Wang, J. X. 2002. *What Every Engineer Should Know about Decision Making under Uncertainty*. Boca Raton, FL: CRC Press.
Wang, J. X. 2005. *Engineering Robust Designs with Six Sigma*. Upper Saddle River, NJ: Prentice Hall.
Wang, J. X. 2010. *Lean Manufacturing: Business Bottom-Line Based.* Boca Raton, FL: CRC Press.
Wang, J. X. 2012. *Green Electronics Manufacturing: Creating Environmental Sensible Products*. Boca Raton, FL: CRC Press.
Wang, J. X., and Roush, M. L. 2000. *What Every Engineer Should Know about Risk Engineering and Management*. Boca Raton, FL: CRC Press.
Wemmerlov, U., and Hyer, N. L. 1989. Cellular manufacturing in the U.S. industry: A survey of users. *International Journal of Production Research,* 27(9), 1511–1530.
Wemmerlov, U., and Johnson, D. J. 1997. Cellular manufacturing at 46 user plants: Implementation experiences and performance improvements. *International Journal of Production Research,* 35(1), 29–49.
Wicks, E. M., and Reasor, R. J. 1999. Designing cellular manufacturing systems with dynamic part populations. *IIE Transactions,* 31, 11–20.

chapter four

Bottom-line-based cellular manufacturing

The purpose of this chapter is to present a method for the system design of a manufacturing cell that aims for profit maximization over a certain period of time. The presented method makes use of simulation, design of experiments, regression analysis, Taguchi methods, and a profit model to generate several feasible and potentially profitable designs from which a decision maker must choose the best alternative based on profit value, robustness, and other practical considerations specific to each case. The application of the method is illustrated with a case study where a partial design of a manufacturing cell is accomplished in a manufacturing company.

4.1 Profitability: Business bottom line

Though very technical and detailed, the concept of cellular production is basically simple: get a finished product from raw materials to shipment as efficiently and as profitably as possible. Cellular manufacturing systems and layouts essentially separate the production line into segments, or cells, sometimes called modules. Each cell, consisting of both workers and production machinery, is dedicated to a particular component of the manufactured product. Ideally, workers and equipment comprising a particular cell are trained and configured to be able to take over the processes of another cell when necessary, thus minimizing downtime and wastage of raw material.

Profitability is the principal goal for most manufacturing companies in the world. In order to reach this goal, managers have to deal with several decision-making situations of a different nature and scope. However, it is not always clear if the decisions taken at the moment will have an impact on profitability.

Technology and cellular manufacturing have combined to streamline the production processes of numerous established and start-up manufacturing facilities worldwide. Lean systems, such as Kaizen and Six Sigma, to name just two, though very often high in start-up cost, provide both a short- and long-term benefit in reducing the waste common to the traditional production line. The bottom line in any manufacturing enterprise is profit. Cellular manufacturing has proven to dramatically increase profits. One example is when a company decides to change the organization of its production systems to cellular manufacturing (CM). Basically, there are two important decisions to make when faced with this situation:

1. The identification of part families and machine groups that constitute each cell
2. The system design of each of the cells

In this chapter, the main purpose is to develop and evaluate a method that deals with the second major decision in CM: the system design of a manufacturing cell. The system design of a manufacturing cell is defined in this context as the decision-making activity of setting control factors in a manufacturing cell at acceptable levels with the objective to minimize or maximize some performance measure. The performance measure to be maximized in this method is the profit of the manufacturing cell. The presented method is divided into four phases:

1. Modeling
2. Experimentation
3. Analysis
4. Evaluation

The method integrates several techniques of analysis such as simulation, design of experiments, regression analysis, Taguchi methods, and the use of a profit model to generate a set of designs. A decision maker must select the best design of the set based on profit, robustness, as well as on other practical considerations specific to each case.

4.2 Economic aspects of cellular manufacturing

Cell formation (CF) is the first phase of cellular manufacturing, and deals with the identification of the part family or families and associated machine groups that constitute each cell (Askin et al., 1997). A complete review of the techniques dedicated to CF is presented in a paper by Selim et al. (1998).

The second phase of CM consists of the system design of each of the previously identified cells. Typical decisions in this phase include equipment layout; selection/design of tooling and fixtures; design of material handling equipment; determination of the number of machine operators; assignment of the operators to the machines or workstations; specification of the capacity of buffers between workstations; and the formulation of a machine-setup policy in a workstation (Park and Lee, 1995). Other factors that have to do with operation and control of the cell are included in this phase since they have proven to have an important influence on the performance of a manufacturing system as suggested by a study presented by Harris (1987). Among these factors, loading, scheduling, dispatching, and sequencing, as well as the order quantities and frequency of replenishment of raw materials, can be included.

Regarding CF, the first phase of CM, Askin and Subramanian (1987) recognized the need for an objective economic basis for the selection of the best cellular configuration. The authors developed an approach for a group technology configuration considering costs of work-in-process and cycle inventory, intragroup material handling, setup, and processing as well as fixed machine costs. The objective of this approach is to determine groups of parts and machines that minimize the costs incurred in the operation of the cellular manufacturing system.

The evaluation of the total cost of groupings of parts and machines in CF was the motivation for the development of a cost model proposed in a paper by Needy et al. (1998). The proposed cost model takes into consideration machine investment, machine setup, and material move cost. Its goal is to evaluate the total cost of a given set of cellular configurations generated by different CF techniques, providing an economic decision criterion for these alternatives. Other examples of economic evaluation in CF can be found in a review by Singh (1993).

Regarding the system design of a manufacturing cell, the second phase in CM, Mosca et al. (1992) addressed a flexible manufacturing system (FMS) dimensioning problem with an approach that integrated stochastic simulation, experimental design, analysis of variance, and response surface analysis. Basically, this approach seeks to generate an FMS design by maximizing production of the system and then evaluating this design with a cost function. The method used by Mosca et al. attempts to minimize costs by introducing combinations that satisfied the production requirements first; however, the conflicting nature of both performance measures, production and cost, makes finding a minimum cost design a difficult task. The compromise between these two performance measures is further complicated by the existence of incommensurable units (pieces and money). This method does not evaluate the profitability of a design, which would make the design selection process simpler by having a single response in terms of money.

In summary, it can be said that there is a need for the creation of models that emphasize the economic aspects of CM in both of its principal phases: cell formation and system design of each of the manufacturing cells. Regarding the second phase, in the system design of each of the manufacturing cells, it is necessary to create a model that allows the carrying out this activity while keeping in mind the potential profit of the cell design. Focusing on only one performance measure, profit, could make the design selection process simple and understandable.

4.3 Incorporating supervisory control and data acquisition (SCADA) into cellular manufacturing

4.3.1 Typical SCADA system design schematic

Suresh (1990) addressed the need to justify the investment of flexible automation systems and proposed a structure for a decision support system (DSS) to evaluate different multimachine system (MMS) configurations physically and economically in a simultaneous way. The DSS structure intends to integrate several analytical, experimental, and rule-of-thumb methods to evaluate these MMS configurations. The financial evaluation takes into account sales revenue, material costs, capital costs, salvage values, depreciation, operating cost, and time value of the assets. However, this evaluation is limited to the combinations of factors the decision maker is able to figure out and introduce to the DSS on a trial-and-error basis.

Figure 4.1 expands on the SCADA system to show a typical SCADA architecture and how the SCADA system interfaces with other enterprise information systems. In addition to the individual remote terminal units (RTUs) that connect with the master terminal unit (MTU) through communications media (i.e., wire, phone, cellular, radio, microwave, or satellite), the MTU also connects with various peripheral devices, as well as with other information systems. Connectivity to other information systems, even those within the enterprise, usually takes place across a firewall or similar barrier to reduce the potential for unauthorized access.

The SCADA system receives real-time data from the RTUs along with scheduling information (demand and supply data) from other information systems. The scheduling information is important for the SCADA system to know what operations are required to meet the scheduled demand and supply requirements.

The SCADA system can operate independent of other enterprise systems, such as accounting, human resources, engineering, and intranets. Indeed, these systems and the SCADA system are better off not being connected. Connecting systems expands the potential accessibility of both systems. SCADA systems, because of their critical nature, should be connected only to the systems with which they require connectivity. A firewall is typically implemented to separate the SCADA system from these other information systems. A firewall is a software device designed to allow certain data to flow from one system to another, while prohibiting any data to flow in the opposite direction. Firewalls, however, are not the end all to information system security. Firewalls have vulnerabilities and can be penetrated if not hardened and protected.

A functioning SCADA system contains a wide variety of hardware components, from computers to RTUs, from telephone modems to satellite

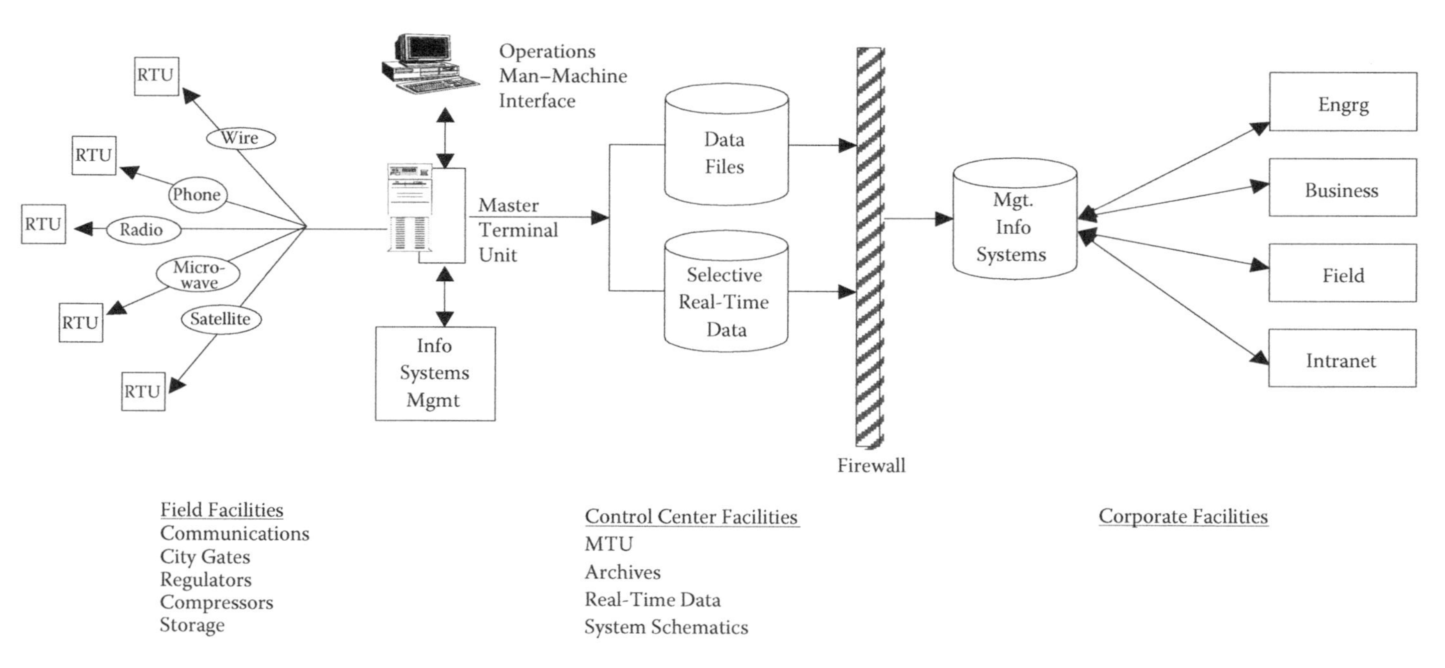

Figure 4.1 Typical SCADA architecture.

radios, from cell phones to microwave towers. It also contains many different software components, from programmable microchips within a field recording device to a powerful SCADA operating program. Designing a SCADA installation for an energy infrastructure system requires that the appropriate components be matched and integrated into the design to obtain the results the operator requires. Few energy operating companies have the in-house technical expertise to design such a system. Contractors known as system integrators design the majority of these systems.

There are more than 200 commercial SCADA equipment manufacturers (both hardware and software) and the number continues to increase. Some of the larger companies offer both hardware and software and if requested will also provide integrator services. There are also many independent integrators that have no tie to any hardware or software vendor but will use any combination that will satisfy the needs of the client.

The current SCADA trend is for companies to purchase off-the-shelf PC-based systems that utilize open protocols and architecture. These systems make the total package design simpler, whether the design is prepared in-house or by an integrator. The pricing of SCADA systems can range from $1,500 to $1.5 million (or higher) depending on the sophistication of the system, the size of the system to be operated, and the selected telecommunications media.

4.3.2 *Desirable features of a SCADA system*

Each SCADA system is unique. Some, particularly those operated by small local utilities, may do little more than gather data and issue reports. These simple systems may have no connection to any of the other enterprise systems operated within the utility. Others, particularly newer systems and those operated by larger entities, may include all the bells and whistles that are currently available. Each operator must decide which features are appropriate and affordable for its system. With a nod to the preceding differences, the more desirable features of a SCADA system are as follows:

- A user-friendly (PC/X-windows/graphics) man–machine interface
- Automatic monitor and control capability
- Supply-and-demand load management
- Off-line processing at designated workstation(s)
- Integrated environments
- Extensive historical data manipulation capability (trend analysis, history matching, output vs. time, output vs. weather, etc.)
- Extensive processing power
- Enhanced data throughput
- Extremely quick response
- Rapid scan frequency (scans/hour)

- Intelligent field units (programmable logic controllers [PLCs], intelligent electronic devices [IEDs])
- On-line and off-line complex network analysis
- Real-time supply/demand side economic calculations
- Automatic system manipulation (e.g., voltage and power factor correction, pressure/flow adjustment)
- Distributed processing power
- Steady-state and transient flow analysis capabilities
- Training simulator

These features all have value to a SCADA operator. Some are of greater or lesser interest depending upon the nature of the particular energy system being operated (electric or gas). Some of these features are costly, and some require substantial computer resources. Others are imperative for modern systems, almost without regard to cost. It is unlikely that many systems will contain all these features in the immediate time frame considering the prohibitive costs. The operator must always decide the extent of capability that he can afford and which compromises must be made.

SCADA systems have become necessary tools for system operation, because they enable the operator to perform the task of system control better and at a lower cost than he can without such a system. Initially, the expected benefits from SCADA were small and were primarily operational in nature. However, the unmistakable economic benefits were seen quickly. These have actually been the driving force behind the proliferation of SCADA throughout the energy industry. The presence of a SCADA system provides both operational and economic benefits. These benefits are explained next.

4.3.3 Operational considerations of incorporating SCADA into cellular manufacturing

- *Geographic area of operation* (size of service area)—Without SCADA capability, control equipment distributed over a large geographic area dictates that key control points must be staffed around the clock because of the need to make timely and potentially frequent adjustments to the distribution system. A SCADA system offers control of these remote facilities from a central control facility without the need for on-site crews except for normal maintenance.
- *Size of company* (number of customers)—The complexity of distribution system operations usually increases with the number of customers, with a concomitant increase in the number of field control actions required to maintain system stability. Without SCADA capability, the number of employees required to complete these actions

also increases. In emergency situations, without centralized oversight, corrective actions in one location may aggravate conditions in an interconnected area. A SCADA system provides essentially real-time access to each of these points from a central control facility, again without the need for on-site crews.

- *Improved timeliness and frequency of operational data*—SCADA systems allow for real-time monitoring of many remote sites and the immediate logging of associated data. Each point can be polled for data many times each hour, as compared to the preautomation method of once or twice per hour, by phone.
- *Integrated operational analysis*—SCADA systems provide graphic schematics of entire systems, including real-time physical data. This enables the operator to observe at a glance the effect of a regional upset on the operation of the system as a whole and aids him in deciding the appropriate countermeasures.
- *Improved precision and accuracy of operational data*—Deregulation has eliminated energy transporters' ability to provide a supply "cushion" for their customers. SCADA systems enable system operators to obtain precise measurement data from remote sites, providing them with the ability to maintain accurate throughput information and make appropriate and timely adjustments, reducing the likelihood of overtake or undertake penalties.
- *Documentation of measurement equipment accuracy*—A SCADA system provides a continuous record of each variable used in calculating delivered energy units, as well as any constants that the system assumes may be used in meeting regulatory requirements, such as U.S. Environmental Protection Agency environmental compliance requirements.
- *Rapid response to emergencies or upsets*—SCADA systems provide audible alarms whenever an operating parameter moves beyond a preset range or even when data is not received. This allows the operator to take prompt action to correct the problem. Many times a situation can be handled from the control center without having to send personnel to the site.

4.3.4 Benefits to business bottom line

- *Reduced personnel requirements*—A SCADA system allows oversight and control of the entire delivery system, including the most remote points, without having personnel stationed at or visiting the sites other than for maintenance purposes. Personnel requirements can be reduced to the level needed for normal maintenance and emergencies, and some physical facilities (buildings, parking lots, associated real estate and vehicles) can be eliminated.

- *Decreased cost of operation*—Decreased operational costs are derived primarily from the reduction in personnel and associated physical facilities and equipment required to monitor and control the distribution system.
- *Improved operational efficiency*—The ability of a SCADA system to deliver real-time data on the operation of the energy delivery systems allows operators to more precisely adjust the systems to optimal conditions and to run them close to operational limits. Such operations can reduce costs by delaying the need for expansions. They also can facilitate entry into new market segments, thereby improving profitability.
- *Improved load-balancing ability*—SCADA capability provides up-to-the-minute information on supply and demand, which minimizes the need for expensive balancing services.

Table 4.1 summarizes the SCADA impact and benefits to energy infrastructures. The majority of these provide positive benefits, either operational or economic in nature.

The reader should recognize that personnel reduction in favor of a SCADA system is a two-edged sword. Although SCADA eliminates the need for substantial personnel numbers, every company operating a

Table 4.1 Impact of a SCADA System on Company Operations

SCADA Impact	SCADA benefit
Reduced personnel requirements	Reduced cost
	Restricted ability to operate in an emergency without a SCADA system
Real-time information, which provides operational flexibility that can be used to meet market demands not previously accessible	Increased income
Better management of commodity movements	Increased efficiency, reduced or delayed expansion needs
Earlier knowledge of problems, facilitating earlier decision making and problem solving	Improved reliability
Availability of timely and precise data on system operation	Improved ability to analyze and predict system operations on a real-time basis
	Improved ability to maximize system deliverability by permitting operations closer to its operational limits (maximum and minimum)

SCADA system assumes that in the event of a system-wide SCADA failure they will temporarily man the key points throughout the system as a stopgap means of obtaining critical data. That is, they will revert to the old operational formula of regular communications by phone. Unfortunately, once they have taken the economic option of force reduction they no longer have adequate numbers to operate manually for anything beyond a few days, especially in times of critical weather. Typically, the short-term solutions available to an operator include calling back personnel who have been transferred to other departments, hiring retirees, and hiring contractors. None of these is adequate for a long-term SCADA outage.

4.4 Reliability challenge of incorporating SCADA systems into cellular manufacturing

Operators of energy infrastructures currently face a challenging operational environment. Working within a framework of increasingly competitive markets, smaller profit margins, mergers, and consolidations, along with a complex regulatory climate, these operators are responsible for making countless daily decisions that affect the safety of the public and the viability of the company. SCADA systems have become a crucial tool for providing strategic, timely, and accurate information to decision makers in a real-time operational environment. As the business environment evolves, SCADA system operations are destined to become even more critical, and therefore more valuable, to company operations.

SCADA systems themselves are also becoming more complex, evolving toward more automated control and additional support to the operators, with the incorporation of artificial intelligence, embedded chips, graphical user interfaces, and databases to provide increasing sophistication to the information they provide to decision makers. User needs have dictated that software components must be more easily customized to provide desired configurations, and that they are based on newer language and design paradigms. PLCs are replacing the RTUs, which provide additional control capabilities and alternative operational approaches.

Every SCADA operator expects the system to fail occasionally, at least in part. Most of these types of failures are nonevents, because they are corrected before anything untoward occurs. Some of these failures correct themselves, such as short-term communications interruptions. In other cases, system redundancy provides a workaround solution. Others are repaired in the field before the situation becomes critical.

For many operators, the communications system is the most vulnerable, and also the least manageable, component of the SCADA system.

SCADA signals passing over the public phone system travel an unseen path. A dig-in somewhere near the control center could conceivably sever all communications between the MTU and the RTUs in the field. More remote hits might disrupt only a small portion of the system. Natural phenomena, such as sunspots, ice and snow, and excessive heat can disrupt wireless transmissions, but usually these will be of short duration. Obviously, signals that are marginal in normal operations are the most susceptible to unfavorable climatic conditions.

As technological sophistication increases at all levels, however, so does exposure to technological intrusion. The possibility that someone (a disgruntled worker, unhappy customer, or terrorist) might interrupt a portion of the communications system is real. The security of the systems and their communication pathways is an issue that still needs to be fully recognized and addressed by the vendors and users of these systems. SCADA systems are replacing field personnel and in several companies the option to operate the system manually no longer exists. Thus, the requirement that the SCADA system operates efficiently, reliably, and securely has never been greater.

At the same time that the energy industry dependence on SCADA is increasing, the threats and vulnerabilities to information technology (which includes SCADA) are also dramatically increasing. Protecting SCADA systems from potential outages due to hardware failures, communication disruptions, and software difficulties is, by itself, a significant challenge. The challenge is increased when internal and external factors that could compromise these systems are considered. SCADA systems inherit the vulnerabilities of computers and networks in addition to SCADA-specific vulnerabilities. By their very nature, SCADA systems are vulnerable because they generally do not employ even common rudimentary information protection measures found in other information technology systems. Most users (the first-line SCADA operators) are skeptical of the potential for cyberintrusion, and find most security protocols to be unnecessary complications to otherwise routine operations. This makes such intrusions more feasible, because those in the best position to prevent it are not alert.

It is not surprising that first-line operators (and upper management, too, in most cases) have not embraced the notion that SCADA systems are vulnerable to cyberintrusion. There is no public history of an intruder-caused SCADA failure, and added complexities in software design appear to be unnecessary and costly obstructions to operations that they would rather see made simpler. Encryption might make it more difficult to hack into a SCADA system, but the additional coding may significantly lengthen the time for each polling sequence, resulting in fewer polls per hour, and therefore less accurate data. Longer polling times also increase the likelihood that a portion of a transmission will be disrupted, resulting in no data at all for that poll.

These concerns have slowed the development of cyberhardened systems because the software vendors, who have the technical resources to develop such systems, have been discouraged by operators from incorporating encryption into current software versions. In addition, software vendors feel economic pressure to bring products to market quickly, with the result that products are sometimes flawed and often do not contain well-designed security interfaces. Information on these vulnerabilities is distributed via hacker sites and chat rooms, detailing what the vulnerabilities are and how to exploit them. Automated exploitation tools have also increased the threat.

Furthermore, it is estimated that industrial and foreign espionage in North America has increased dramatically in recent years, and it has been acknowledged by the U.S. government that other countries have established nationally sponsored information warfare efforts that are targeted against North American commerce. Many of these adversaries possess the sophistication necessary to target SCADA systems. Clearly, anti-intrusion measures for all facets of cybertechnology are needed to protect the nation's energy infrastructure.

In recent years, the energy infrastructure has become very dependent on SCADA, and thus on the information technology backbone on which SCADA depends. However, adequate protection processes for these systems do not exist. The growing popularity of Internet-based SCADA systems presents a significant risk since numerous hackers attempt to break into systems every day.

Currenty, there is an effort underway, which is supported by various energy industry associations and entities, to develop a viable encryption standard for use in SCADA and other cyberdependent applications. Preliminary work indicates that the cost of such a protocol will be very reasonable for new software, with a negligible impact on system speed or capability. There will be some degradation in speed for certain retrofit options.

4.4.1 The impact of SCADA failures

It is difficult to quantify in any general way the impact of SCADA failures on energy infrastructures. Each energy utility employs a unique operational philosophy when adding SCADA to its overall system operations. This operational philosophy determines the precise relationship between the SCADA system and the energy system being controlled. This relationship determines the impact on operations when a SCADA failure occurs. This section examines the factors and range of SCADA failure impacts. The effect on a company with a SCADA system failure depends on a number of factors, which include those listed next.

- *Geographic spread*—A system that is widely dispersed geographically may be impacted more severely by a SCADA failure. This is due to the additional time required to get personnel to appropriate sites, as well as the potentially greater personnel requirements to secure a widespread system.
- *Number of RTUs*—Large numbers of critical RTUs require large staffs for physical operations in the event of a wide-scale SCADA system failure.
- *System complexity*—More complex energy systems are usually more reliant on SCADA capabilities, and therefore more vulnerable in the event of a SCADA system failure.
- *Design philosophy*—Fail-safe features such as maximum pressure restricting valves or excess flow valves can be independent of the SCADA system. Safeguards such as these can mitigate a SCADA failure because they maintain system viability within preset maximum and minimum values, which can be modified on a seasonal basis.
- *Type of operation (distribution or transmission)*—Transmission pipeline systems generally depend more on SCADA capabilities than distribution systems because of their widespread and geographically disperse areas of operation.
- *Time of day*—The time of day is a factor from the perspective of a failure that occurs during the morning or evening demand peaks. A SCADA failure of an energy system during peak times would have a higher impact.
- *Day of week*—The day of the week is a factor because mobilization of crews can be more problematic and time consuming on weekends and holidays.
- *Interdependencies*—Infrastructures are becoming increasingly interdependent. For example, electric utility systems require natural gas to provide gas-fired generators. A SCADA failure in the natural gas system could cascade to causing problems in the electric power system. Conversely, a power failure of an electric utility system will cause a failure in a natural gas SCADA system if it lasts longer than the battery backup built into the system. And all SCADA systems depend on a communications infrastructure, whether it is a private system or commercially operated.

4.4.2 Types of SCADA failures

There are three main types of SCADA system failures. Each type of failure will have an impact on operations, even though the impact may not be significant overall. The three types of SCADA system failures include:

1. *Loss of data from a single point or region of the system*—A localized SCADA failure can cause a loss of data transfer that may or may not impact system operations. SCADA systems are designed to handle communications, hardware, and software failures. During periods of bad weather, for example, it is common to have communications problems between RTUs and the MTU. A system will generally continue to operate on either its last SCADA setting or at a system default value. In most SCADA systems, there are many RTUs that provide data to the MTU but cannot be adjusted remotely. Such a point will continue to operate normally whether it is in contact with the MTU. A problem occurs when a controllable RTU setting needs to be changed and cannot be because of the SCADA failure.
2. *System controller blind*—The SCADA system provides the system controller with important real-time information feeds on the system. If a problem occurs at the MTU the controller is no longer obtaining this crucial information. If the system is not under stress and the current RTU settings are appropriate for current system operations, this may not cause a problem in the short term. If the loss of data continues for several hours or past the end of the accounting day, there will be accounting and financial confusion even if operations are not stressed. If the system is under peak conditions or stressed from physical problems, a total SCADA failure may cause a problem even if it is short term.
3. *Controller intervention is disrupted*—If a SCADA failure precludes a system controller from intervening in operating the system through RTU settings, this may cause a problem. If the system is stable, not under stress, and limited changes are occurring to the system state, then little or no impact from the SCADA failure may occur. If, however, the system is under stress, or the system is changing and RTU set points need to be updated but cannot be because of the SCADA failure, then impacts may occur.

Perhaps the greatest determinant of the potential impact caused by a SCADA failure is the design philosophy employed when the SCADA system is built. Some of the alternatives are discussed next.

SCADA systems can be designed to be totally passive. A passive system observes conditions and collects data on a predetermined regular schedule. It can also forward commands entered by an operator that will adjust physical settings in the field. This type of system is quiescent except when commanded to act and is not essential to the operation of the energy system. Just as the energy system operated with frequent (or infrequent) input from its human controllers before SCADA, so, too, will it operate whether or not SCADA is watching. If system conditions change, however, manual input is required, whether it can be done in the control center or must involve on-site action by field personnel.

Some SCADA systems are more directly involved in controlling system operations. These systems are programmed to observe and collect data, just as are the passive systems. These systems will also make field adjustments when commanded. However, as they continue to monitor the device, they will make appropriate adjustments in order to maintain the setting originally entered by the operator. This can be done in more than one way. The MTU can be programmed to transmit a corrective signal whenever it observes the value moving beyond its normal range. This can increase the traffic on the communications system and can only be done half as often as the system scans that point. If the SCADA system fails, this system becomes identical to the passive system described in the previous paragraph.

A more efficient method employs intelligent RTUs in the field. The RTU will receive the electronic command from the MTU, just as in the first scenario. The RTU then monitors its own performance without oversight from the MTU. It can do this almost instantaneously, without tying up the communications system. It will maintain the precise setting until it receives new instructions from the MTU. If the SCADA system fails, the RTU will continue to hold the desired value, even though system conditions may degrade. Although no system can provide output greater than input, this system offers a higher level of operational security than the first two.

The worst-case scenario is a SCADA failure that leads to a system shutdown. Although most SCADA systems contain fail-safe protections, under extreme conditions insertion of field personnel may not be quick enough or effective enough to avoid a system shutdown following a SCADA failure. Because SCADA systems have become crucial to system operations, system operators must be alert to any conditions that threaten any portion of SCADA operations. When extreme conditions threaten, many companies elect to man key facilities that might otherwise be difficult to reach quickly in an emergency.

4.5 Total productive maintenance

Total productive maintenance (TPM) seeks to engage all levels and functions in an organization to maximize the overall effectiveness of production equipment. In addition, this method tunes up existing processes and equipment by reducing mistakes and accidents. Whereas maintenance departments are the traditional center of preventive maintenance programs, TPM seeks to involve workers in all departments and levels, from the plant floor to senior executives, to ensure effective equipment operation.

Autonomous maintenance, a key aspect of TPM, trains and focuses workers to take care of the equipment and machines with which they work. TPM addresses the entire production system lifecycle and builds a solid, plant floor-based system to prevent accidents, defects, and breakdowns. TPM focuses on preventing breakdowns (preventive maintenance), "mistake-proofing" equipment (or poka-yoke) to eliminate product defects and nonconformance, or to make maintenance easier (corrective maintenance), designing and installing equipment that needs little or no maintenance (maintenance prevention), and quickly repairing equipment after breakdowns occur (breakdown maintenance).

The goal is the total elimination of all losses, including breakdowns, equipment setup and adjustment losses, idling and minor stoppages, reduced speed, defects and rework, spills and process upset conditions, and start-up and yield losses. The ultimate goals of TPM are zero equipment breakdowns and zero product defects, which lead to improved utilization of production assets and plant capacity.

SCADA systems have become crucial to the operations of the energy infrastructures because they provide valuable and timely data, promote efficient operations, and are cost-effective. For these reasons the use of SCADA technology continues to increase in all facets of the energy industry. Additionally, water, wastewater, and railroad infrastructures utilize SCADA technology as well. Many of the commercially available SCADA control packages can be used in any of these applications.

Systems similar to SCADA systems are routinely seen in factories, treatment plants, and so forth. These are often referred to as distributed control systems (DCSs). They have similar functions to SCADA systems, but the field data gathering or control units are usually located within a more confined area. Communications may be via a local area network (LAN), and will normally be reliable and high speed. However, many municipally operated water and wastewater SCADA systems connect with facilities several miles distant and utilize the public Internet for telecommunications. Such systems are more vulnerable to outside intrusion than are those using LANs or other local communication systems.

SCADA systems are subject to all the vulnerabilities inherent in information systems. These include vulnerabilities built into its RTUs, its MTUs, and its software and hardware. They also inherit the vulnerabilities inhabiting every other information system with which they interconnect, as well as the operating systems under which they run. A hacker can penetrate a SCADA system through its interconnection with another information system. Firewalls provide only limited protection in many cases.

The communications infrastructure is a significant vulnerability, both as a carrier for SCADA signals and as a carrier for fallback operations via telephone. It is highly visible, and is a target for attack in its own right, not just as a SCADA component. A SCADA operator can reduce its exposure to communications shutdown if it can add redundant communications paths to its SCADA system.

SCADA systems can fail due to hardware problems, software problems, or communications problems. All three must be functional for the system to operate. The effects of a SCADA failure range from no impacts to system shutdown. The impact to a specific system vary depending on geography, number of RTUs, system complexity and type of system, operational state, time, and interdependencies.

An operating infrastructure and its SCADA system should be designed as an integrated system. All elements of the system that provide an aspect of control, both electronic and mechanical, must be compatible in order to provide the most stable, fail-safe operation.

As SCADA systems continue to replace field personnel and meet the needs of real-time market requirements, the operational security of SCADA systems will continue to grow in importance. The data gathered by a SCADA system may represent electrical values, such as voltage or current flow within an electrical distribution. Or it may represent pressure or temperature, such as within a gas or water distribution system. Whatever the original form, a measured value must be converted to a binary signal that can be transmitted through a standard communications network and deconverted at the other end.

In a pressurized system, such as a gas transmission line, an electronic pressure transducer is used to measure pressure. This instrument senses the pressure within the system and compares the value to a preestablished range that is programmed into the transducer. It then converts the value to a comparable portion of its 4 to 20 ma output range. The resulting current signal is an analog for the system pressure.

Bibliography

Agarwal, A., and Sarkis, J. 1998. A review and analysis of comparative performance studies on functional and cellular manufacturing layouts. *Computers and Industrial Engineering*, 34, 77–98.

Askin, R. G., and Subramanian, S. P. 1987. A cost-based heuristic for group technology configuration. *International Journal of Production Research*, 25, 101–113.

Askin, R. G., Selim, H. M., and Vakharia, A. J. 1997. A methodology for designing flexible cellular manufacturing systems. *IIE Transactions*, 29, 599–610.

Atwater, J. B., and Chakravorty, S. 1995. Using the theory of constraints to guide the implementation of quality improvement projects in manufacturing operations. *International Journal of Production Research*, 25, 1737–1760.

Buchwald, S. 1994. Throughput accounting: A recipe for success. *1994 Conference Proceedings* of *American Production and Inventory Control Society*, pp. 635–637.

Byrne, D. M., and Taguchi, S. 1987. Taguchi approach to parameter design. *Quality Progress*, 20, 19–26.

Cary, D. F. 1993. What the big boys really think about you: How cycle time reduction shows up in the financial analysis. *Conference Proceedings American Production and Inventory Control Society*, pp. 704–707.

Connor, W. S., and Zelen, M. 1959. *Fractional Factorial Experimental Designs for Factors at Three Levels*. Washington, D.C.: National Bureau of Standards.
Dean, B. R., Kaye, M., and Hand, S. 1997. A cost-based strategy for assessing improvements in manufacturing processes. *International Journal of Production Research*, 35, 955–968.
Evren, R. 1987. Interactive compromise programming. *Journal of the Operational Research Society*, 38, 163–172.
Fowlkes, W. Y., and Creveling, C. M. 1995. *Engineering Methods for Robust Product Design*. New York: Addison-Wesley.
Fry, T. D. 1995. Japanese manufacturing performance criteria. *International Journal of Production Research*, 33, 933–954.
Gershon, M. 1984. The role of weights and scales in the application of multiobjective decision making. *European Journal of Operational Research*, 15, 244–250.
Goicoechea, A., Hansen, D. R., and Duckstein, L. 1982. *Multiobjective Decision Analysis with Engineering and Business Applications*. New York: Wiley & Sons.
Goldratt, E. M., and Cox, J. 1984. *The Goal: A Process of Ongoing Improvement*. Great Barrington, MA: North River Press.
Harris, C. R. 1987. Modeling the impact of design, tactical, and operational factors on manufacturing system performance. *International Journal of Production Research*, 35, 479–499.
Keeney, R. L., and Raiffa, H. 1976. *Decisions with Multiple Objectives*. New York: Wiley & Sons.
Kelton, D. W., Sadowski, R. P., and Sadowski, D. A. 1998. *Simulation with Arena*. New York: McGraw-Hill.
Lee, S., and Jung, H. J. 1989. A multiobjective production planning model in a flexible manufacturing environment. *International Journal of Production Research*, 27, 1981–1992.
Lewis, H. S., Sweigart, J. R., and Markland, R. E. 1996. An interactive decision framework for multiple objective production planning. *International Journal of Production Research*, 34, 3145–3164.
Liggett, H. R., and Trevino, J. 1992. The application of multi-attribute techniques in the development of performance measurement and evaluation models for cellular manufacturing. *1992 Conference Proceedings, Flexible Automation and Information Management*, pp. 711–722.
Montgomery, D. C. 1997. *Design and Analysis of Experiments*. New York: John Wiley & Sons.
Mosca, R., Giribone, P., and Drago, A. 1992. A simulation approach to FMS dimensioning including economic evaluations. *Flexible Automation and Information Management, 1992*, pp. 771–781.
Needy, K. L., Billo, R. E., and Colosimo Warner, R. L. 1998. A cost model for the evaluation of alternative cellular manufacturing configurations. *Computers and Industrial Engineering*, 34, 119–134.
Park, H., and Lee, T. 1995. Design of a manufacturing cell in consideration of multiple objective performance measures. In *Planning, Design, and Analysis of Cellular Manufacturing Systems,* edited by A. K. Kamrani, H. R. Parsaei, and D. H. Liles. New York: Elsevier.
Raffish, N. 1994. Activity-based costing—Part II. *1994 Conference Proceedings of American Production and Inventory Control Society*, pp. 631–633.

Romero, C., Tamiz, M., and Jones, D. F. 1998. Goal programming, compromise programming, and reference point method formulations: Linkages and utility interpretations. *Journal of the Operational Research Society*, 49, 986–991.
Sarfaraz, A. R., and Emamizadeh, B. 1993. Product costing for the concurrent engineering environment. *1993 Conference Proceedings of American Production and Inventory Control Society*, pp. 688–690.
Selim, H. M., Askin, R. G., and Vakharia, A. J. 1998. Cell formation in group technology: Review, evaluation and directions for future research. *Computers and Industrial Engineering*, 14, 3–20.
Shi, Y., and Yu, P. L. 1989. Goal setting and compromise solutions. In *Multiple Criteria Decision Making and Risk Analysis Using Microcomputers*, edited by B. Karpak and S. Zionts. Berlin, Germany: Springer-Verlag.
Singh, N. 1993. Design of cellular manufacturing systems: An invited review. *European Journal of Operational Research*, 69, 284–291.
Stevens, M. E. 1993. Activity-based costing/management (ABC/ABM): The strategic weapon of choice in the global war competition. *1993 Conference Proceedings of American Production and Inventory Control Society*, pp. 699–703.
Suresh, N. 1990. Towards an integrated evaluation of flexible automation investments. *International Journal of Production Research*, 28, 1657–1672.
Wang, J. X. 2002. *What Every Engineer Should Know about Decision Making under Uncertainty*. Boca Raton, FL: CRC Press.
Wang, J. X. 2005. *Engineering Robust Designs with Six Sigma*. Upper Saddle River, NJ: Prentice Hall.
Wang, J. X. 2010. *Lean Manufacturing: Business Bottom-Line Based*. Boca Raton, FL: CRC Press.
Wang, J. X. 2012. *Green Electronics Manufacturing: Creating Environmental Sensible Products*. Boca Raton, FL: CRC Press.
Wang, J. X., and Roush, M. L. 2000. *What Every Engineer Should Know about Risk Engineering and Management*. Boca Raton, FL: CRC Press.
Wemmerlov, U., and Hyer, N. L. 1989. Cellular manufacturing in the U.S. industry: A survey of users. *International Journal of Production Research*, 27, 1511–1530.
Wemmerlov, U., and Johnson, D. J. 1997. Cellular manufacturing at 46 user plants: Implementation experiences and performance improvements. *International Journal of Production Research*, 35, 29–49.
Wizdo, A. 1993. A methodology for relevant costing and profitability. *1993 Conference Proceedings of American Production and Inventory Control Society*, pp. 694–698.
Yu, P. L. 1985. *Multiple-Criteria Decision Making: Concepts, Techniques, and Extensions*. New York: Plenum Press.
Zionts, S. 1989. Multiple criteria mathematical programming: An updated overview and several approaches. In *Multiple Criteria Decision Making and Risk Analysis Using Microcomputers*, edited by B. Karpak and S. Zionts. Berlin, Germany: Springer-Verlag.

chapter five

Performance evaluation of cellular manufacturing systems

In this chapter, the method presented is comprised of four sequential phases: modeling, experimentation, analysis, and evaluation. The modeling phase is necessary to summarize and integrate information on how the manufacturing cell is intended to work. The experimentation phase guides the search for relationships between the controllable factors, the noncontrollable factors, and the profitability of their combination. The analysis phase allows the user to establish and use these relationships to generate different solutions. Finally, the evaluation phase entails the identification and confirmation of the most profitable solutions, as well as the selection of a final design for the manufacturing cell.

5.1 Performance evaluation: Cellular manufacturing

Cellular manufacturing (CM) is used to overcome the deficiencies of job-shop manufacturing, including excessive setup times and high levels of in-process inventories. But their performance superiority is found to diminish when the demand becomes unstable. Adaptive cellular systems are especially designed for dynamic demands. Here, the performance of an adaptive cellular system is compared to a system using process layout, using simulation, when they execute production in a multiperiod dynamic demand environment. The adaptive design performs comparatively better than a process layout in terms of reduced work-in-process (WIP) inventory and manufacturing lead time. The managing of semifinished parts at the end of a period and the effort to relocate machines after every period are disadvantages of adaptive cellular systems.

In cellular manufacturing, part families are identified and machine cells are formed such that one or more part families can be fully processed within a single machine cell. A machine–part matrix that represents the processing requirements of parts in the product mix is the main source of data for the development of a cellular manufacturing system. As a result, product mix variations affect the structure of the machine–part matrix and the performance of the corresponding cellular manufacturing system. A cellular manufacturing system (CMS) is designed to perform best for a specific product mix. When parts in the product mix or their production volume changes, the performance of the cellular manufacturing will change as well. Since in real-world situations product mix variation is inevitable and it is not feasible to modify the cellular manufacturing system to fit the changes, it is necessary to determine the sensitivity of the performance of the cellular manufacturing system to such changes. To do so, the performance of the cellular manufacturing system should be evaluated under a range of possible compositions of parts in the product mix.

Generally, a cellular manufacturing system is designed based on a single machine–part matrix. Since the cellular manufacturing system is designed to best fit the processing requirements of the original machine–part matrix, the new composition of parts in the machine–part matrix may adversely affect the system performance. Although it is desirable to modify the cellular manufacturing system to meet the new processing requirements, for all practical purposes such a modification is not feasible. Therefore, it is necessary to foresee the changes in the product mix and to evaluate the performance of the cellular manufacturing system subject to these changes.

As shown in Figure 5.1, the presented method is comprised of four sequential phases:

1. *Modeling*—The modeling phase is necessary to summarize and integrate information on how the manufacturing cell is intended to work.
2. *Experimentation*—The experimentation phase guides the search for relationships between the controllable factors, the noncontrollable factors, and the profitability of their combination.
3. *Analysis*—The analysis phase allows the user to establish and use these relationships to generate different solutions.
4. *Evaluation*—The evaluation phase entails the identification and confirmation of the most profitable solutions, as well as the selection of a final design for the manufacturing cell.

As a machine–part matrix deviates from its original structure, it is expected that the performance of the corresponding cellular manufacturing system deteriorates. Consequently, it is necessary to evaluate the performance of the cellular manufacturing system under these changes to determine the relationship between the percentage of changes of parts in the product mix and the rate of change in the performance. Such an evaluation is crucial in the decision making regarding the conversion from job shop to cellular manufacturing. Since the performance evaluation of any manufacturing system, particularly a cellular manufacturing system, is complex, simulation modeling offers the best solution to the performance evaluation problem. Mean flow time and WIP inventories are widely used performance measures in the evaluation of

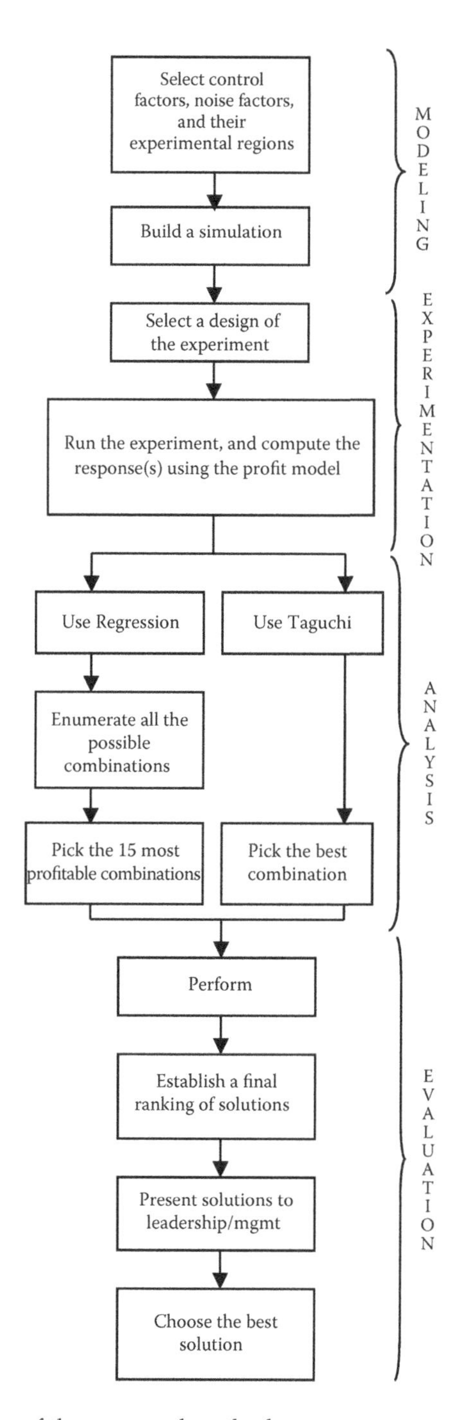

Figure 5.1 Diagram of the proposed method.

cellular manufacturing systems. These two performance measures will be employed in this chapter.

5.2 Modeling cellular manufacturing performance

The first phase of the method, modeling, represents an opportunity to screen, summarize, and integrate diverse information on how the manufacturing cell is intended to work. A cellular manufacturing system belongs to a family of modern production methods, which many industrial sectors have used beneficially in recent years. In fact, this system is an application of group technology (GT) determining cell formation (i.e., clustering part families and machine grouping) and layout design (i.e., intercell and intracell layouts). During the last two decades, a number of researchers have carried out scientific studies on static production and deterministic demand states. However, in the real world, a CM model often consists of a large number of variables and constraints. To extract a solution from such problems requires a large amount of computer time, memory, and processing power by current optimization software packages.

By building a simulation model, the people involved in the design of the cell are able to develop intuition on the interrelationships of different factors of the manufacturing system being studied. The objective in this phase is to build a simulation model that allows the inclusion of specific factors, controllable and noncontrollable, that are likely to have an impact in the profitability of the cell for their study. The controllable factors of a system are classified as control factors, and the noncontrollable factors as noise factors throughout this study. These two terms were adopted from the Taguchi methods (Fowlkes and Creveling, 1995).

The design of cellular manufacturing systems involves many structural and operational issues. One of the important design steps is the formation of part families and machine cells. In this chapter, a comprehensive mathematical model for the design of cellular manufacturing systems based on tooling requirements of the parts and the tooling that is available on the machines is presented. The model incorporates dynamic cell configuration, alternative routings, lot splitting, sequence of operations, multiple units of identical machines, machine capacity, workload balancing among cells, operation cost, cost of subcontracting part processing, tool consumption cost, setup cost, cell size limits, and machine adjacency constraints. Numerical examples are presented to demonstrate the model and its potential benefits.

Control factors are the independent variables x_1, x_2, ..., x_k that have influence on a dependent variable y (response) and that have the capability to be adjusted with the objective to minimize or maximize this response.

In the modeling phase, it is necessary to define the control factors that are likely to have an impact on the profitability of the system. These will be varied in the experiment. It is also necessary to specify the ranges over which these factors will be varied (the experimental region) as well as the specific levels at which runs will be made.

The selection of control factors must be performed by using the experience and the engineering intuition of the design team. The task consists of determining if a particular control factor has an impact on the production levels, the inventory levels, the operation expenses, or the equipment expenses of the manufacturing cell. If it does, that control factor is very likely to affect the profitability of the system and should be included in the study.

The control factors need to be discrete when using the proposed method. For example: 2 workers, 4 machines, 32 pieces, and so on. It is recommended to use three equally spaced levels for each factor for two reasons:

1. To cover a wide experimental region.
2. To provide the ability to investigate quadratic effects in the response of the experiment.

Noise factors are variables that can influence the performance of a system and that are not under our control (Fowlkes and Creveling, 1995). In general, noise factors are included in an experiment to seek robustness (insensitivity to variation) in a design.

The noise factors should also be selected based on the experience and the engineering intuition of the design team. Its influence on the production levels, the inventory levels, the operation expenses, or the equipment expenses of the manufacturing cell makes a noise factor a candidate to be included in the study due to its potential effect on the profitability of the system.

It is recommended that noise factors be kept at two levels since the objective of their inclusion is to induce a fair amount of variation to investigate robustness in the design and not to determine their optimal levels. Once the factors to be included in the study are defined, it is necessary to build a computer simulation model of the manufacturing cell. The simulation model should allow the design team to easily experiment with the previously defined control and noise factors, as well as to compute statistics on specific aspects of interest. For the purposes of the proposed method, it is necessary for the simulation model to keep statistics on machine utilization per machine, as well as production, scrap, and WIP inventory per part type.

5.3 Experimental design for cellular manufacturing performance evolution

5.3.1 Axiomatic design: Establishing a cellular manufacturing system's design structure

The second step of the method, experimentation, involves the choice of an experimental design to run the simulation model and the actual run of the experiment. Cellular manufacturing has been recognized as an efficient and effective way to improve productivity in a factory. In recent years, there have been continuous research efforts to study different facets of CMS. They are developed to satisfy only one or limited functional requirements (FRs) of the CMS design. The literature does not contain much published research on CM design, which includes all design aspects. In this section, we provide a framework for the complete CMS design. It combines axiomatic design (AD) and experimental design (ED) to generate several feasible and potentially profitable designs. The AD approach is used as the basis for establishing a systematic CMS design structure. ED has been a very useful tool to design and analyze complicated industrial design problems. AD helps secure valid input factors to the ED. An element of the proposed framework is demonstrated through a numerical example for a cell formation problem with an alternative process.

This section provides a framework and a road map for people who are ready to transform their traditional production system from process orientation to cellular orientation, based on AD principles. A feedback mechanism for continuous improvement is also suggested for evaluating and improving the cellular design against preselected performance criteria. A complete implementation of the proposed methodology at a manufacturing company and resulting performance improvements are also provided.

New product development is one of the most powerful but difficult activities in business. It is also a very important factor affecting final product quality. There are many techniques available for new product development. Experimental design is now regarded as one of the most significant techniques. In this section, we will discuss how to use the technique of experimental design in developing a new product—an extrusion press. In order to provide a better understanding of this specific process, a brief description of the extrusion press is presented. To ensure the successful development of the extrusion press, customer requirements and expectations were obtained by detailed market research. The critical and noncritical factors affecting the performance of the extrusion press were identified in preliminary experiments. Through conducting single-factorial experiments, the critical factorial levels were determined. The relationships between the performance indexes of the extrusion press and the four critical factors were determined on the basis of multifactorial experiments. The mathematical models for the performance of the extrusion press were established according to a central composite rotatable design. The best combination of the four critical factors and the optimum performance indexes were determined by optimum design. The results were verified by conducting a confirmatory experiment. Finally, a number of conclusions became evident.

In today's highly competitive world, customer needs are becoming more important than ever before. This new paradigm brought forth the importance of AD principles, which brings customer needs to the forefront in system reengineering processes. The ultimate goal of AD is to establish a scientific basis for design and to improve design activities by providing the designer with a theoretical foundation based on logical and rational thought processes and tools. In accomplishing this goal, the AD provides a systematic search process through the design space to minimize the random search process and determine the best design solution among many alternatives.

In the initial concept design stage of a manufacturing system, there is often a great deal of uncertainty about objectives that it must meet. Knowledge of data such as the number of product variations, future customer demand, and other production planning requirements are often estimates at this stage that are subject to frequent revision. The application of analytical tools at this stage is premature and such tools are often ill-suited to the volatility and incompleteness in manufacturing data. To deal with the lack of formal methods at this design stage, a top-down, functional design approach has been developed to guide the design of the manufacturing system from the initial statement of high-level business objectives to their later translation into lower level, more detailed design requirements (Cochran, 1999). A structured approach based on an axiomatic design (Suh, 1990) allows functional requirements to be mapped to design parameters (DPs), which in manufacturing system design correspond to subsystems such as cells and equipment. Furthermore, it provides an understanding of the many interdependencies that arise between these subsystems. Also, the manufacturing system design decomposition (Figure 5.2) provides a means to relate high-level objectives to the design parameters of the subsystems.

As an example of a subsystem design, an equipment design approach (Arinez and Cochran, 1999) has been proposed that offers the procedural and quantitative link to requirements from the manufacturing system decomposition. The approach consists of the four major steps shown in the center of Figure 5.2. The first step deals with identifying the set of system requirements (blacked out in Figure 5.2) that influence equipment design from the decomposition. Second, this set must be transformed in a manner (view creation) that permits them to be understood by the various parties involved in the design of equipment. The third step is the generation

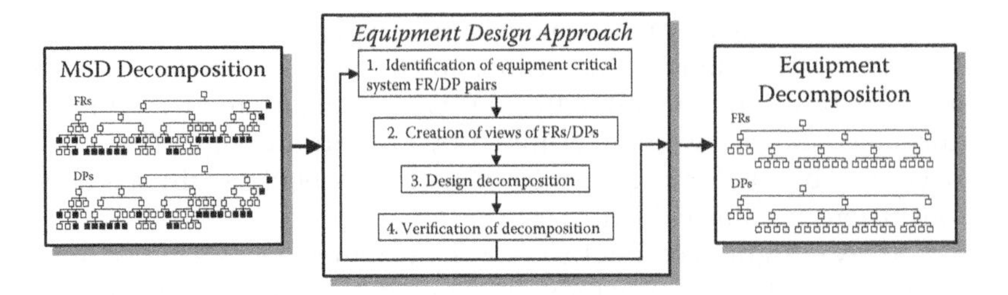

Figure 5.2 Equipment design approach based on manufacturing system design decomposition. (Adapted from Fowlkes, W. Y., and Creveling, C. M., 1995, *Engineering Methods for Robust Product Design*, New York: Addison-Wesley.)

of the equipment design decomposition from these requirements (far right in Figure 5.2). The final step is the validation of the resultant equipment design on the basis of whether it meets systems requirements and its effect on the performance of the manufacturing system design.

The design of the CMS starts following the preliminary stage. At this stage, the AD approach to cellular manufacturing is presented to the transition and design team in a systematic and scientifically sound order. These guidelines, which are developed based on the independence axiom, provide necessary steps in transforming an existing process-oriented system into a CMS. With this systematic approach the designer will be guided by a methodology for appropriate analysis techniques.

Transition to cellular manufacturing follows after all cellular manufacturing steps are successfully completed. At this stage, the production is achieved through CMS. The system is now generating necessary databases and information for comparing the system performance with set target goals on some business metrics. Based on target values and achievements, new target values are established and appropriate system modifications and changes are affected through CMS improvement principles provided in our proposed procedure. These principles are also based on AD principles. This continuous feedback and improvement principles are also in agreement with the spirit of Lean manufacturing and Kaizen activities.

5.3.2 Experimental design: Generate potentially profitable designs

An experimental design dictates the number of runs, the levels at which the factors must be set on each run, and the sequence in which these runs must be performed. In summary, the experimental design dictates how to run the experiment. Referring to Figure 5.2, a layout of the experimental design required for the use of the proposed method consists of six parts:

1. Control factor array
2. Noise factor array

3. Set of responses
4. Column for the mean of the responses
5. Column for the standard deviation of the responses
6. Column for the signal-to-noise (S/N) ratios

A diagram showing the placement of each of these parts is shown in Figure 5.2. The control factor array selected should allow the designer(s) to estimate the full quadratic model in Equation (5.1):

$$y = \beta_0 + \sum_{i=1}^{k} \beta_i x_i + \sum_{i=1}^{k} \beta_{ii} x_i^2 + \sum_{i=1}^{k-1} \sum_{j>i}^{k} \beta_{ij} x_i x_j + \varepsilon \tag{5.1}$$

where
x_i = control factor *i*
x_j = control factor *j*
β_i = regression coefficient *i*
β_{ii} = regression coefficient *ii*
β_{ij} = regression coefficient *ij*
ik = number of control factors
ε = random error

In column (1) of Table 5.1, a 3^k full factorial design (where k is the number of control factors) or a 3^{k-p} fractional factorial design should be used to fit the responses (profit values) of the experiment. A collection of 3^{k-p} fractional factorial designs can be found in Connor and Zelen (1959). For the noise factor array, column (2), it is recommended to choose an orthogonal array.

The design of experiments (DOE) is used to plan the simulation experiments. The performance of each of the three layouts is analyzed statistically by means of operational parameters such as machine utilization, throughput, average distance traveled by parts, and average WIP. The results from the simulation experiments indicate that the performance of virtual cellular manufacturing falls between that of functional and cellular manufacturing. Also, we find that the performance of a virtual cellular layout is often relatively superior to that of a functional layout and marginally inferior to a cellular layout.

DOE has been a very useful tool to design and analyze complicated industrial design problems. The one shown in Figure 5.2 is an L4 orthogonal array, where the number 4 refers to the number of runs to be performed. An L4 orthogonal array can accommodate up to three noise factors at two levels each. Taguchi methods practitioners use this type of array extensively. A collection of available orthogonal arrays can be found in the book by Fowlkes and Creveling (1995). The sets of responses in each

Table 5.1 Layout of the Experimental Design

(2) Noise Factor Array				
I	LOW	HIGH	HIGH	LOW
J	LOW	HIGH	LOW	HIGH
K	LOW	LOW	HIGH	HIGH

	(1)								(3)				(4)	(5)	(6)
Run	Control factor array								Set of responses				Mean	Standard	S/N_{LTB}
Number	A	B	C	D	E	F	G	H	1	2	3	4	Values	Deviation	Ratio
1	1	1	1	1	1	1	1	1	y_{11}	y_{12}	y_{13}	y_{14}	$\bar{y}_1$	S_1	S/N_1
2	1	1	2	2	2	2	2	2	…	…	…	…	…	…	…
3	1	1	3	3	3	3	3	3	…	…	…	…	…	…	…
4	1	2	1	1	2	2	3	3	…	…	…	…	…	…	…
5	1	2	2	2	3	3	1	1	…	…	…	…	…	…	…
6	1	2	3	3	1	1	2	2	…	…	…	…	…	…	…
7	1	3	1	2	1	3	2	3	…	…	…	…	…	…	…
8	1	3	2	3	2	1	3	1	…	…	…	…	…	…	…
9	1	3	3	1	3	2	1	2	…	…	…	…	…	…	…
10	2	1	1	3	3	2	2	1	…	…	…	…	…	…	…
11	2	1	2	1	1	3	3	2	…	…	…	…	…	…	…
12	2	1	3	2	2	1	1	3	…	…	…	…	…	…	…
13	2	2	1	2	3	1	3	2	…	…	…	…	…	…	…
14	2	2	2	3	1	2	1	3	…	…	…	…	…	…	…
15	2	2	3	1	2	3	2	1	…	…	…	…	…	…	…
16	2	3	1	3	2	3	1	2	…	…	…	…	…	…	…
17	2	3	2	1	3	1	2	3	…	…	…	…	…	…	…
18	2	3	3	2	1	2	3	1	y_{181}	y_{182}	y_{183}	y_{184}	$\bar{y}_{18}$	S_{18}	S/N_{18}

row (column 3) comprise the response values resulting from a combination of control factors across each combination of noise factors. The computation of the response values will be explained in the following section, where the profit model is discussed.

Column (4) will store the mean values of the responses averaged across the noise factor array in each row. The values in column (6) are the values of the standard deviation (S_i) of the response values calculated across the noise factor array. This value is a measure of the variability of the sample data. Finally, the column for the S/N ratios, column (6), will be used for storing the ratios under the larger-the-better (LTB) scenario computed using Equation (5.2). The S/N_{LTB} ratios are used when the desired value of the response is the largest possible number.

$$S/N_{LTB_i} = -10\log\left[\frac{1}{n}\sum_{i=1}^{n}\left(\frac{1}{y_{ij}^2}\right)\right] \tag{5.2}$$

where

iy_{ij} = individual response value from *i*th combination of control factors and *j*th combination of noise factors

n = total number of combinations of noise factors for each combination of control factors

The use of the S/N_{LTB} ratio is restricted to response values that are continuous nonnegative numbers ranging from 0 to infinity. If this is not the case, transformation of these values will be necessary to make the set of responses adequate for the use of the S/N_{LTB} ratio. The steps for this transformation are as follows:

1. Make a copy of the set of responses (a new set of responses).
2. Identify the smallest negative response in the new set.
3. Multiply that response times –1.
4. Add the value obtained to all the responses in the new set.
5. Compute the S/N_{LTB} ratio for the new set of responses.

This data manipulation will create two different copies of the set of responses. During the analysis phase, the original set of responses will be used for the regression analysis, and the new set will be used for the Taguchi methods analysis. The S/N ratio is a measure that reflects the variability in the response of a system caused by noise factors. S/N ratios are widely used as part of the Taguchi methods (Fowlkes and Creveling, 1995, pp. 53–120). In summary, the experimentation phase of the proposed method entails three basic activities:

1. Determining an experimental design that accommodates the control and noise factors selected previously to run the experiment.
2. Running the experiment by following the experimental design.
3. Computing the necessary data from the collected response values for further analysis.

The computation of the response values will be explained in detail in the next section.

5.3.3 Developing a profit model for cellular manufacturing

The response values discussed in the previous section are computed with the profit model defined by Equation (5.3):

$$\text{PROFIT} = \sum_i s_i \left[P_i^f - \sum_j P_j^r r_{ji} \right] - \alpha \sum_j P_j^r I_j - \sum_m h_m (e_m + \tau_m) - lwa - c \tag{5.3}$$

where

- s_i = number of finished sold products of type i produced during the analysis period
- P_i^f = price of finished product type i
- r_{ji} = amount of raw material type j used to produce 1 unit of product type i
- P_j^r = purchasing price of raw material type j per measuring unit
- α = holding cost per analysis period
- I_j = average inventory level of raw material type j
- h_m = number of operation hours of machine m during the analysis period
- e_m = cost of the energy consumed by machine type m per unit time
- τ_m = tooling cost for machine m per unit time
- l = length of the analysis period in hours
- w = average wage rate per hour for an associate
- a = number of associates
- c = equivalent uniform analysis period cost of equipment

The variable s_i is taken directly from the statistics collected during the simulation. This variable is expressed in terms of physical units, and it comprises the products that have already been sold, that is those that were produced either to meet a customer's order or to meet the demand for the planning period. P_i^f is the price at which the company sells the products being produced by the manufacturing cell. The amount of raw material used to produce one finished product, r_{ji}, takes into account the scrap level of the manufacturing cell. To compute this ratio, the number of units of raw material consumed by the cell (composed of the figures for production and scrap obtained from the simulation) is divided by the number of pieces produced by the cell.

The purchasing price of the raw material, P_j^r, is the unitary price at which the supplier sells the raw material to the company. The holding cost per analysis period, α, is the cost of having inventory for a period of time.

The holding cost is computed by multiplying an interest rate (properly compounded for the analysis period) and the price of the raw material, P_j^r.

The average inventory level, I_j, refers to the average WIP inventory computed through the simulation. The number of operation hours per machine, h_m, is computed by multiplying the utilization of a specific machine times the length of the analysis period in hours (l). The energy cost, e_m, is the cost of all the energy consumed by a specific machine per hour. Energy, in the context of this method, is any consumable, other than raw material and tooling, essential for a machine to perform its processing functions. It includes, among other things, gas, oil, and electricity.

The tooling cost of a machine, τ_m, includes the cost of the different tools that need to be replaced with certain frequency in a machine to ensure a product is within a customer's specifications. This cost needs to be expressed as a cost per hour. Variables l, w, and a refer to the length of the analysis period in hours, the average wage rate per hour, and the number of associates employed in the manufacturing cell, respectively.

Finally, variable c represents the equivalent uniform analysis period cost of the equipment. The computation of c is similar to the computation of the equivalent uniform annual cost (EUAC), where a sum of money is converted to its equivalent annual cash flow. Four factors must be taken into account for the computation of the equivalent uniform analysis period cost of the equipment: installed cost of the equipment, interest rate, expected life of the equipment, and salvage value. The formula to compute c is defined by Equation (5.4):

$$c = C_I(A/P,i,n) - V_s(A/F,i,n) \tag{5.4}$$

where

C_I = installed cost of the machine
i = interest rate for the analysis period
n = number of analysis periods that equal the expected life of the machine
V_s = salvage value

5.4 *Analyzing cellular manufacturing performance*

The third step of the method is analysis, its objective is to establish the relationships between the control factors and responses based on the data obtained in the experimentation phase, and use these relationships to generate different alternative designs. Two analysis techniques will be used in this phase: regression methods and Taguchi methods. The details of such techniques will be explained in the following sections.

5.4.1 Regression methods

A manufacturing cell consists of a cluster of functionally dissimilar machines or processes that are placed in close proximity to one another and are dedicated to the manufacture of a set of part families. Parts are grouped into part families depending on the similarity in parts' geometry, manufacturing processes required, or both. The objective of regression analysis is to identify several activities as variables describing the major implementation process. The ultimate goal is to determine which activities have the greatest impact on the degree of success. Best subsets of the variables (the ones that have the most impact on CM success) are generated using exploratory descriptive (rather than predictive) regression models. The results may provide a preliminary identification of best practices for CM implementation.

The application of regression analysis in this method's objective is to relate the responses obtained in the experimentation phase with the control factor array. Regression analysis involves fitting the quadratic model in Equation (5.1) to predict the profit (response) at various combinations of the control factors.

Regression analysis may be done by using a statistics software package. These packages typically use the method of least squares to estimate the regression coefficients. The method of least squares chooses β's in Equation (5.1) so that the sum of the squares of the error terms (ε) is minimized.

In this step of the method it is necessary to obtain the quadratic regression model based on the control factor array and the column of the averages. The model must then be coded in a spreadsheet to predict the profit of all the possible combinations of control factors (an enumeration of the full factorial). The predicted profit values for all the possible combinations will serve as the basis for sorting the combinations in decreasing order (the most profitable design will be shown on top of the list). The 15 most profitable designs will be selected to proceed with the method to the final phase—evaluation.

5.4.2 Taguchi methods

The cellular manufacturing problem has captured the attention of many manufacturers and researchers. Although a considerable amount of literature exists in this domain, most of the research focuses on a single-criterion cell configuration problem, for example, to maximize the throughput. However, to achieve overall efficiency and to balance different aspects of shop performance, management needs multiple performance measures. In this study, we use flow time, waiting time, and WIP as the performance criteria. In order to robustly and optimally design and control the cellular manufacturing system, we modify both the Taguchi method and the response surface methodology (RSM) to the multicriteria situation. The Taguchi method provides robust design, while RSM seeks optimal design. A combination of both methods helps the cellular manufacturing system achieve its fullest potential and therefore find robust design and operating conditions that maximize system efficiency.

Taguchi method analysis finds the combination of control factors that has the lowest variation across the combinations of noise factors, that is, it aims for robustness in a solution. It makes use of the S/N_{LTB} ratio defined by Equation (5.3). The S/N_{LTB} is used to find the most robust combination and the largest possible response simultaneously. The inclusion of Taguchi methods in the proposed method has the objective to provide a baseline design with desirable characteristics (robustness and high response value) to compare the designs generated by regression analysis.

CM is an important application of group technology that can be used to enhance both flexibility and efficiency in today's small-to-medium lot production environment. The crucial step in the design of a CM system is the cell formation (CF) problem, which involves grouping parts into families and machines into cells. The CF problem is increasingly complicated if parts are assigned with alternative routings (known as generalized group technology problem). In most of the previous works, the route selection problem and CF problem were formulated in a single model, which is not practical for solving large-scale problems. It is suggested that a better solution could be obtained by formulating and solving them separately in two different problems. The objective is to apply Taguchi methods for the route selection problem as an optimization technique to get back to the simple CF problem, which can be solved by any of the numerous CF procedures. At this point, the S/N_{LTB} ratios may be listed in the respective column (6) in the experimental design (see Table 5.1), and it must have been verified that the responses (in the set of responses) are nonnegative numbers to ensure the reliability of the analysis. If not all positive the data must be adjusted as described previously.

To generate a cell design using Taguchi methods analysis it is necessary to create plots of the control factor effects for the S/N ratio and for the mean response. The optimum set points of the control factor levels are the ones at which the S/N ratio is maximized. This "pick-the-winner" set point selection is done in a factor-by-factor basis by visual inspection of the plots. Finally, if the plots for the S/N ratio show ties then plots of the control factor effects on the mean response must be also considered to break these ties and generate a design.

5.5 Evaluation

The fourth phase of the method is evaluation, its objective is to select the best design from the alternatives generated in the analysis phase. For this purpose, it is necessary to perform confirmation runs, interact with the decision maker to include practical issues not included in the study, and finally, to let the decision maker decide which manufacturing cell design is best for the intended application. In addition, the main effect of each part and analysis of variance (ANOVA) are introduced as a sensitivity analysis aspect that has been completely ignored in previous research.

The 15 most profitable designs generated by regression analysis, as well as the design generated by the Taguchi methods analysis, are run in the simulation model created in the modeling phase. These are called confirmation runs. It is necessary to run these designs to get the response values (profit) across the noise factor array, and compute the average and the standard deviation of each design. The confirmation runs allow for establishing a final rank of the designs in terms of their profitability and to give an idea of the expected variation across the noise factor array.

The information generated from the confirmation runs must be presented to the decision maker for their evaluation. Usually, decision makers have other considerations of practical matters that can affect the decision on selecting a final design. Among these considerations, one could find budget limitations, staffing policies, constraints in the integration of the manufacturing cell with the rest of the plant, space limitations, and so forth. At the end, the best design will be selected based not only on profitability, but also on robustness, and compliance with the case-specific practical considerations.

5.6 How to build a secure networked cellular manufacturing system: Starting from a gap analysis

Although there is currently a fair amount of effort being dedicated to reducing and mitigating the risks of electronic attacks to electric power systems, there is still much to be done. Some of these gaps include:

- Lack of a comprehensive approach to security that includes policies and procedures that apply to all devices, connections, and likely scenarios
- Lack of communication and cooperation between the utility's information technology (IT) department and supervisory control and data acquisition (SCADA) engineers and operators
- Lack of regular assessments of SCADA vulnerabilities
- Lack of ways of performing real-world tests on systems without risking interruption of service to customers
- Lack of security devices for low-bandwidth, real-time control networks
- Lack of common testing methods for identifying vulnerabilities
- Lack of methods to quantify cybersecurity risks

- Lack of reliable, inexpensive sensors and devices to ensure physical security of substations
- Quantity of operating system and application security patches makes staying current a difficult task
- Bottlenecks in the power grid make widespread outages more likely

The following are some potential solutions to help eliminate these gaps:

- Develop security methods and devices that work well with low bandwidth communications channels in real-time control networks
- Implement better security in substation devices (e.g., remote terminal units [RTUs], intelligent electronic devices [IEDs])
- Develop testing criteria for control systems vulnerabilities
- Perform real-world testing on live or test systems to isolate and mitigate vulnerabilities of integrated systems
- Implement cybersecurity in the most widely used SCADA protocols (e.g., distributed network protocol [DNP], Inter-Control Center Communications Protocol [ICCP])
- Streamline patch management of control systems
- Reduce potential consequences of cyberattacks by strengthening the power grid
- Implement security requirements that apply to all systems and components critical to the operation, maintenance, and continued reliability of the electrical power grids
- Intelligence on configuration of potential adversarial systems
- Development of smart systems (alert and block [defense] or alter attack based on configuration [offense])

5.7 Just-in-time

Just-in-time production, or JIT, and cellular manufacturing are closely related, as a cellular production layout is typically a prerequisite for achieving JIT. JIT leverages the cellular manufacturing layout to significantly reduce inventory and WIP. JIT enables a company to produce the products its customers want, when they want them, in the amount they want.

Under conventional mass production approaches, large quantities of identical products are produced, and then stored until ordered by a customer. JIT techniques work to level production, spreading production evenly over time to foster a smooth flow between processes. Varying the mix of products produced on a single line, sometimes referred to as "shish-kebab production," provides an effective means for producing the desired production mix in a smooth manner.

JIT frequently relies on the use of physical inventory control cues (or kanban) to signal the need to move raw materials or produce new components from the previous process. In some cases, a limited number of reusable containers are used as kanban, ensuring that only what is needed gets produced. Many companies implementing Lean production systems are also requiring suppliers to deliver components using JIT. The company signals its suppliers, using computers or delivery of empty, reusable containers, to supply more of a particular component when they are needed. The end result is typically a significant reduction in waste associated with unnecessary inventory, WIP, and overproduction.

Bibliography

Agarwal, A., and Sarkis, J. 1998. A review and analysis of comparative performance studies on functional and cellular manufacturing layouts. *Computers and Industrial Engineering*, 34, 77–98.

Arinez, J. F., and Cochran, D. S. 1999. Application of a production system design framework to equipment design. *32nd CIRP International Seminar on Manufacturing Systems*.

Askin, R. G., and Subramanian, S. P. 1987. A cost-based heuristic for group technology configuration. *International Journal of Production Research*, 25, 101–113.

Askin, R. G., Selim, H. M., and Vakharia, A. J. 1987. A methodology for designing flexible cellular manufacturing systems. *IIE Transactions*, 29, 599–610.

Atwater, J. B., and Chakravorty, S. 1995. Using the theory of constraints to guide the implementation of quality improvement projects in manufacturing operations. *International Journal of Production Research*, 25, 1737–1760.

Black, J. T. 1991. *The Design of the Factory with a Future*. New York: McGraw-Hill.

Buchwald, S. 1994. Throughput accounting: A recipe for success. *1994 Conference Proceedings of American Production and Inventory Control Society*, pp. 635–637.

Byrne, D. M., and Taguchi, S. 1987. Taguchi approach to parameter design. *Quality Progress*, 20, 19–26.

Cary, D. F. 1993. What the big boys really think about you: How cycle time reduction shows up in the financial analysis. *1993 Conference Proceedings of American Production and Inventory Control Society*, pp. 704–707.

Cochran, D. 1999. The production system design and deployment framework. SAE Technical paper 1999-01-1644, *SAE IAM-99 Conference*.

Connor, W. S., and Zelen, M. 1959. *Fractional Factorial Experimental Designs for Factors at Three Levels*. Washington, D.C.: National Bureau of Standards.

Dean, B. R., Kaye, M., and Hand, S. 1997. A cost-based strategy for assessing improvements in manufacturing processes. *International Journal of Production Research*, 35, 955–968.

Evren, R. 1987. Interactive compromise programming. *Journal of the Operational Research Society*, 38, 163–172.

Fowlkes, W. Y., and Creveling, C. M. 1995. *Engineering Methods for Robust Product Design*. New York: Addison-Wesley.

Fry, T. D. 1995. Japanese manufacturing performance criteria. *International Journal of Production Research*, 33, 933–954.

Gershon, M. 1984. The role of weights and scales in the application of multiobjective decision making. *European Journal of Operational Research*, 15, 244–250.

Goicoechea, A., Hansen, D. R., and Duckstein, L. 1982. *Multiobjective Decision Analysis with Engineering and Business Applications*. New York: Wiley & Sons.

Goldratt, E. M., and Cox, J. 1984. *The Goal: A Process of Ongoing Improvement*. Great Barrington, MA: North River Press.

Harris, C. R. 1997. Modeling the impact of design, tactical, and operational factors on manufacturing system performance. *International Journal of Production Research*, 35, 479–499.

Keeney, R. L., and Raiffa, H. 1976. *Decisions with Multiple Objectives*. New York: Wiley & Sons.

Kelton, D. W., Sadowski, R. P., and Sadowski, D. A. 1998. *Simulation with Arena*. New York: McGraw-Hill.

Lee, S., and Jung, H. J. 1989. A multiobjective production planning model in a flexible manufacturing environment. *International Journal of Production Research*, 27, 1981–1992.

Lewis, H. S., Sweigart, J. R., and Markland, R. E. 1996. An interactive decision framework for multiple objective production planning. *International Journal of Production Research*, 34, 3145–3164.

Liggett, H. R., and Trevino, J. 1992. The application of multi-attribute techniques in the development of performance measurement and evaluation models for cellular manufacturing. *1992 Conference Proceedings of Flexible Automation and Information Management*, pp. 711–722.

Monden, Y. 1993. *Toyota Production System*. Norcross, GA: Industrial Engineering and Management Press.

Montgomery, D. C. 1997. *Design and Analysis of Experiments*. New York: John Wiley & Sons.

Mosca, R., Giribone, P., and Drago, A. 1992. A simulation approach to FMS dimensioning including economic evaluations. *1992 Conference Proceedings of Flexible Automation and Information Management*, pp. 771–781.

Needy, K. L., Billo, R. E., and Colosimo Warner, R. L. 1998. A cost model for the evaluation of alternative cellular manufacturing configurations. *Computers and Industrial Engineering*, 34, 119–134.

Park, H., and Lee, T. 1995. Design of a manufacturing cell in consideration of multiple objective performance measures. In *Planning, Design, and Analysis of Cellular Manufacturing Systems*, edited by A. K. Kamrani, H. R. Parsaei, and D. H. Liles. New York: Elsevier.

Raffish, N. 1994. Activity-based costing—Part II. *1994 Conference Proceedings of American Production and Inventory Control Society*, pp. 631–633.

Romero, C., Tamiz, M., and Jones, D. F. 1998. Goal programming, compromise programming, and reference point method formulations: Linkages and utility interpretations. *Journal of the Operational Research Society*, 49, 986–991.

Sarfaraz, A. R., and Emamizadeh, B. 1993. Product costing for the concurrent engineering environment. *1993 Conference Proceedings of American Production and Inventory Control Society*, pp. 688–690.

Selim, H. M., Askin, R. G., and Vakharia, A. J. 1998. Cell formation in group technology: Review, evaluation and directions for future research. *Computers and Industrial Engineering*, 14, 3–20.

Shi, Y., and Yu, P. L. 1989. Goal setting and compromise solutions. In *Multiple Criteria Decision Making and Risk Analysis Using Microcomputers*, edited by B. Karpak and S. Zionts. Berlin, Germany: Springer-Verlag.

Singh, N. 1993. Design of cellular manufacturing systems: An invited review. *European Journal of Operational Research*, 69, 284–291.

Stevens, M. E. 1993. Activity-based costing/management (ABC/ABM): The strategic weapon of choice in the global war competition. *1993 Conference Proceedings of American Production and Inventory Control Society*, pp. 699–703.

Suh, N. P. 1990. *Principles of Design*. New York: Oxford University Press.

Suresh, N. 1990. Towards an integrated evaluation of flexible automation investments. *International Journal of Production Research*, 28, 1657–1672.

Wang, J. X. 2002. *What Every Engineer Should Know about Decision Making under Uncertainty*. Boca Raton, FL: CRC Press.

Wang, J. X. 2005. *Engineering Robust Designs with Six Sigma*. Upper Saddle River, NJ: Prentice Hall.

Wang, J. X. 2010. *Lean Manufacturing: Business Bottom-Line Based*. Boca Raton, FL: CRC Press.

Wang, J. X. 2012. *Green Electronics Manufacturing: Creating Environmental Sensible Products*. Boca Raton, FL: CRC Press.

Wang, J. X., and Roush, M. L. 2000. *What Every Engineer Should Know about Risk Engineering and Management*. Boca Raton, FL: CRC Press.

Wemmerlov, U., and Hyer, N. L. 1989. Cellular manufacturing in the U.S. industry: A survey of users. *International Journal of Production Research*, 27, 1511–1530.

Wemmerlov, U., and Johnson, D. J. 1997. Cellular manufacturing at 46 user plants: Implementation experiences and performance improvements. *International Journal of Production Research*, 35, 29–49.

Wizdo, A. 1993. A methodology for relevant costing and profitability. *1993 Conference Proceedings of American Production and Inventory Control Society*, pp. 694–698.

Womack, J. P., Roos, D., and Jones, D. T. 1990. *The Machine That Changed the World*. New York: Rawson Associates.

Yu, P. L. 1985. *Multiple-Criteria Decision Making: Concepts, Techniques, and Extensions*. New York: Plenum Press.

Zionts, S. 1989. Multiple criteria mathematical programming: An updated overview and several approaches. In *Multiple Criteria Decision Making and Risk Analysis Using Microcomputers*, edited by B. Karpak and S. Zionts. Berlin, Germany: Springer-Verlag.

chapter six

Robust design of cellular manufacturing systems

As changing conditions prevail in the manufacturing environment, the design of cellular manufacturing systems, which involves the formation of part families and machine cells, is difficult. This is due to the fact that the machines need to be relocated as per the requirements if adaptive designs are used. The purpose of this chapter is to illustrate the use of robust design with a real application. Some background of the problem is presented first, then each of the steps of the method is illustrated along with the data used throughout the study, and finally results are discussed.

6.1 Robust design versus adaptive design

Cellular manufacturing (CM) is a model for workplace design and has become an integral part of Lean manufacturing systems. Cellular manufacturing is based upon the principles of group technology (GT), which seeks to take full advantage of the similarity between parts, through standardization and common processing. In functional manufacturing, similar machines are placed close together (e.g., lathes, millers, drills, etc.). Functional layouts are more robust to machine breakdowns, have common jigs and fixtures in the same area, and support high levels of demarcation. In cellular manufacturing systems (CMS), machines are grouped together according to the families of parts produced. The major advantage is that material flow is significantly improved, which reduces the distance traveled by materials, inventory, and cumulative lead times. Cellular manufacturing employs setup reduction and gives workers the tools to operate multiple processes that are multifunctional, owning quality improvements, waste reduction, and simple machine maintenance. This allows workers to easily self-balance within the cell while reducing lead times, resulting in the ability for companies to manufacture high-quality products at a low cost, on time, and in a flexible way.

The advantages of CMS are well established. Setups are reduced because each cell handles similar parts. Ideally, all the processing of a part is done within the cell to which it belongs and therefore the shop floor control is better (Needy et al., 1998). The group of workers who work within a cell have a better opportunity for learning all the manufacturing requirements of that cell. This enables them to do activities such as scheduling of jobs and routine maintenance of machines. As a result, their morale and job satisfaction improves (Wemmerlov and Hyer, 1989; Wemmerlov and Johnson, 1997).

But the part families and machine cells of a CMS are designed for a set of parts, product mix, and demand volume. Therefore, when demand changes either in volume or product mix or both, the performance of CMS deteriorates (Morris and Tersine, 1990; Seifoddini and Djassemi, 1997). Many designs of CMS that are suitable for dynamic demands have been suggested in the literature. But there are few performance studies.

The performances of two such designs that are suitable for multiperiod demands are studied here. In multiperiod demand, the demand may vary when the entire plan horizon is considered but is steady for certain subdivisions of the plan horizon. These subdivisions are the periods of steadiness of the dynamic demand. The first of the two designs is the adaptive cellular layout proposed by Wicks and Reasor (1999). In this design, for each period of the multiperiod plan horizon, a cell configuration and part family allocation that is suitable for the demand of that period is used. For successive periods, the part families are reassigned and the machine cells are composed afresh, by means of relocating existing machines and acquiring new ones, whichever is necessary. The cells, their layouts, and part families for all periods are planned at the beginning of the plan horizon itself with an objective of minimizing the total cost of machine acquisition, machine relocation, and material handling for the whole plan horizon. The second design is the robust cellular layout proposed by Pillai and Subbarao (2008). Here, machine relocation as in adaptive layouts is avoided and a layout that will be optimum for all periods is used for the whole plan horizon, but at the same time allowing different part family grouping and assignment that is suitable for each period. The optimization is done with the objective of minimizing the total cost of machine acquisition and material handling for the whole plan horizon.

Simulation is one of the best methods for analyzing complex systems such as manufacturing systems (Ang and Willey, 1984; Flynn and Jacobs, 1986). In simulation, different parameters of the system can be varied to study their effect. This avoids the time and cost of experimenting with real systems. Simulation also helps to compare alternative system designs, in an effective manner, by providing the same set of external environments for the different designs.

The performance of process layout, adaptive cellular layout, and robust cellular layout under dynamic demand have been analyzed using simulation. Performance measures include:

- Manufacturing lead time (MLT)
- Average work-in-process (WIP) inventory
- Production rate

These performance measures are used for the performance study. CMS design strategies to deal with the dynamic production requirements can be grouped under any one of the following two strategies:

- Robust design strategy
- Adaptive design strategy

6.1.1 *Robust design strategy*

A robust design strategy is to design a cellular manufacturing system that is good for the entire planning horizon even though it may not be optimal in any period. The underlying concept of a robust design strategy is to avoid the relocation and the associated disadvantages by forming fixed cells at the start of the planning horizon such that the cells should be capable of producing every period's part families. This strategy generally works with a forecast of product mix and demand changes from period to period of a planning horizon and does not allow the composition of machine cells to change over time. The major drawback of the robust design strategy is that the formation of fixed cells may not be optimal in any period.

6.1.2 *Adaptive design strategy*

The purpose of an adaptive design strategy, on the other hand, is to design a cellular manufacturing system that responds to changing product mix or demand in future periods by rearranging the current manufacturing system. Since the formed cells in the current period may not be optimal for the next period, the reconfiguration of the cells is required. The relocation of cells is associated with cost, which is known as the relocation cost. However, it may be noted that the relocation cost can be offset by the reduced material handling cost achieved through reconfiguration. Therefore, an efficient adaptive design strategy based solution methodology is required for addressing the dynamic CMS design problem.

6.2 *Develop a problem statement*

The performance of two cellular manufacturing systems specially designed to meet dynamic demands, namely, adaptive cellular system and robust cellular system, are analyzed and compared with that of a well-designed process layout. A multiperiod demand is given, where the product mix as well as the demand volume, changes from period to period. A process layout, an adaptive cellular layout, and a robust cellular layout are designed for this demand. The performance of the three manufacturing systems while executing production according to this demand is studied and their performance is compared.

Table 6.1 Comparison of Products Included in the Study

Product	Weight (lb.)	Mix participation
A	40	61%
B	31	18%
C	26	17%
D	48	3%

This study was conducted based on the plans of a company to install a new manufacturing cell for engine components. The company, Progressive Diesel (PD), manufactures components for the transportation industry in a make-to-order fashion. The new manufacturing cell is expected to replace at least three existing dedicated production lines in PD's machining operations department. The scope of this study includes four product classes (A, B, C, and D) that differ mostly in size and production mix participation. A comparison of these products is presented in Table 6.1.

The process for manufacturing products A through D is a four-step sequence:

1. Cold rolling
2. Machining
3. Inspection
4. Marking

A flow diagram of the process in the manufacturing cell is shown in Figure 6.1.

The demand for the different product classes can be represented in a 6 weeks' time frame based on PD's work plan and weekly schedule. The representation of the demand is shown in Table 6.2. The opportunity areas in the design of the manufacturing cell at PD included setting optimal levels for the number of workers, order quantities, frequency, and size of safety stock of raw materials. Decisions concerning the optimal number of machines required to perform the second operation of a given set of parts as well as the optimal number of setups per planning period were included in the study to illustrate the use of the proposed method.

In summary, the present case study involves setting the staffing levels, number of computer numerical control (CNC) lathes for the second operation, as well as delivery quantities, frequencies, and safety stock sizes of raw material at optimal levels to maximize the profit of the new manufacturing cell at PD.

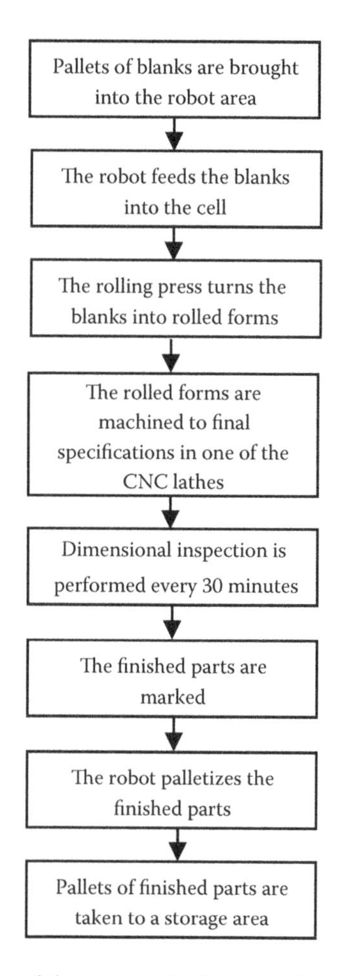

Figure 6.1 Flow diagram of the process in the manufacturing cell.

Table 6.2 Demand per Product Type in a 6-Week Period (Pieces)

Product	Week 1	Week 2	Week 3	Week 4	Week 5	Week 6
A	5200	5200	5200	5200	5200	5200
B	3100	3100	3100			
C				2900	2900	2900
D			810	810		

6.3 Robust design modeling

The first step of the robust design method involves the selection of the factors that will be included in the study. These factors can be either controllable factors or uncontrollable factors. The terminology used with Taguchi methods identifies the former as control factors and the latter as noise factors. These terms will be used throughout this study.

Control factors are the independent variables x_1, x_2, …, x_k that have influence on a dependent variable y (response) and that have the capability to be adjusted with the goal to optimize this response. In general, it is necessary to define the control factors that will be varied in the experiment, the ranges over which these factors will be varied (experimental region), and the specific levels at which runs will be made.

There are nine control factors that were identified by the author in his consultation with the design team. These factors and their levels are explained in the following.

- The first control factor, the number of associates, refers to the number of workers to staff in the cell.
- Control factors 2, 3, and 4, frequency of blank replenishment of product classes A, B, and C, refer to the number of times per week blanks (raw material) of each product must be delivered to the manufacturing cell. Product D was not included in the analysis because of its low participation in the mix (only 3%).
- Control factors 5, 6, and 7, safety stock for product classes A, B, and C, refer to the size of the safety stock that must be kept in-plant with the objective to protect the cell operation against starvation phenomena. The safety stock units were defined in multiples of the estimated number of pieces the machine in charge of the first operation (cold rolling) is able to produce in one shift. This number is estimated to be 480 pieces per shift for any part type.
- The eighth control factor, the number of machines, refers to the number of CNC lathes that must be purchased for the second operation of the parts. The decision to vary the number of machines in the second operation comes from the assumption that the machine for the first operation is considerably more expensive than any of the machines required for the second operation.
- Finally, the ninth factor, the number of setups per 6 weeks, refers to the number of setups or changeovers the cell will have per analysis period (6 weeks) as a result of the decisions made on the schedule for the operation of the cell. This factor was included with the objective to investigate its impact on the profitability of the cell, as well as to establish the objective to achieve the smallest number of setups.

Table 6.3 Control Factors

Control factors	Level 1	Level 2	Level 3
x_1: Number of associates	1	2	3
x_2: Frequency of blanks replenishment product class A	Once/week	Twice/week	Three times/week
x_3: Frequency of blanks replenishment product class B	Once/week	Twice/week	Three times/week
x_4: Frequency of blanks replenishment product class C	Once/week	Twice/week	Three times/week
x_5: Safety stock product class A	2 shifts of production	4 shifts of production	6 shifts of production
x_6: Safety stock product class B	2 shifts of production	4 shifts of production	6 shifts of production
x_7: Safety stock product class C	2 shifts of production	4 shifts of production	6 shifts of production
x_8: Number of machines	2 CNC lathes	3 CNC lathes	4 CNC lathes
x_9: Number of setups in 6 weeks	8	11	14

Table 6.3 shows a summary of the control factors that are to be included in this study regarding the system design of a manufacturing cell.

Noise factors are variables that can influence the performance of a system and that are not under our control (Fowlkes and Creveling, 1995). In general, noise factors are included in an experiment to seek robustness (insensitivity to variation) in a design. Eight factors at two levels each were included in this study. The noise factors were taken from the terms of the profit model in Chapter 5 (Equation 5.3) and are described in the paragraphs that follow. The inclusion of noise factors in this study is for illustration purposes only since no information is available from PD to determine the experimental region for each noise factor.

- The first noise factor, price of raw material of type j, P_r^j, refers to the set of prices of the blanks used to produce each of the product classes A, B, C, and D. If there were estimates available for this noise factor, it would be recommended to use those as low prices, and then determine the high prices by adding a specific increment. Prices of raw materials are more likely to increase than to decrease.
- The second noise factor, interest rate, refers to the interest rate used to compute the inventory holding cost (α) and the equivalent uniform analysis period cost of the equipment (c). If an interest rate is

known, it is recommended to use it as a lower bound and add an increment to set an upper bound with the objective of determining the experimental region.

- The third noise factor, energy cost (e), refers to the cost of the energy consumed by each of the machines in one hour in the cell.
- The fourth noise factor, tooling cost (τ), refers to the cost of a tooling change for each of the machines in the cell. It is assumed that the cold roller and each of the CNC lathes require one tool change per shift and that these machines are the only ones in the cell that require tooling change. The actual figure used in the computations will be in terms of cost per hour.
- The fifth noise factor, machine installed cost, refers to the cost of one CNC lathe, which is the machine required for the second operation of every product produced in the cell.
- If known values are available for the third, fourth, and fifth noise factors, it is recommended to use those as lower bounds, and then introduce an increment to determine an upper bound. Costs are more likely to increase than decrease in real life.
- The sixth noise factor, expected life, refers to the expected life used to compute the equivalent uniform analysis period cost (c) of the machines in the cell. All the machines are assumed to have the same expected life in this study. The U.S Internal Revenue Service (IRS) provides information on the estimation of the expected life of different types of assets.
- The seventh noise factor, salvage value, refers to the future value at the end of the expected life of the CNC lathes. This noise factor was defined as a function of the value of the asset in this case since both noise factors, expected life and salvage cost, are highly related in real life.
- Finally, the eighth noise factor, wage, refers to the hourly wage paid to the associates. Again, if an hourly wage is known it is recommended to use it as a lower bound and then add an increment to determine an upper bound. Wages are also more likely to increase than to decrease.

The noise factors included in the study as well as their experimental regions are summarized in Table 6.4.

Once the factors to be included in the study are defined, it is necessary to build a simulation model of the intended manufacturing cell. This simulation model should allow the designer(s) to easily experiment with the previously defined control and noise factors, as well as to gather statistics on specific aspects of interest.

Building a simulation model at this point in the method allows the designer(s) to integrate a significant amount of information to represent

Table 6.4 Noise Factors

Noise factor	Level 1 (LOW)	Level 2 (HIGH)
Price of raw material of type j (P_r^j) for products (A, B, C, D)	($20, $16, $13, $24)	($32, $26, $21, $39)
Interest rate	20%	30%
Energy cost (e) per machine	$10/hr	$20/hr
Tooling cost (τ) per machine	$15/tooling change	$25/tooling change
Machine Installed Cost	$250,000	$500,000
Expected life	8 years	12 years
Salvage value	1/10 of its cost	1/4 of its cost
Wage (w)	$20/hr	$30/hr

the manufacturing system, bringing clarity and understanding of how the different pieces of information fit together. The simulation model developed for this study is a representation of the manufacturing cell for which the process was described in Figure 6.1. For modeling purposes, the demand of products A, B, C, and D is assumed to follow the pattern described in Table 6.2. This demand pattern takes into consideration the most likely weekly production mixes in a 6-week period with highly likely production quantities.

The control factor array chosen for the present case study was required to accommodate nine control factors (x_1 through x_9 in Table 6.3) at three levels each. The full factorial of this combination (3^9) would have required up to 19,683 runs. Since that amount of runs would have required a considerable amount of simulation time, it was decided to run a fractional factorial (3^{9-p}) with the right size to be able to estimate the 55 regression coefficients in Equation (6.1):

$$y = \beta_0 + \sum_{i=1}^{9} \beta_i x_i + \sum_{i=1}^{9} \beta_{ii} x_i^2 + \sum_{i=1}^{8} \sum_{j>i}^{9} \beta_{ij} x_i x_j + \varepsilon \tag{6.1}$$

A 3^{9-5} experimental design was finally chosen for the control factor array. This array needs 81 runs, and has 81 – 1 = 80 degrees of freedom, which are many more than the 55 degrees of freedom needed for the regression coefficients.

The noise control array was required to accommodate eight noise factors (1 through 8 in Table 6.4). The orthogonal array closest to the requirements of the study was the L12, which can contain up to 11 factors.

The set of responses was computed by coding Table 6.4 in a spreadsheet. Sale prices of products A, B, C, and D were assumed to be $90, $80, $70, and $100 per unit, respectively, and the cost of the machines other than the CNC lathes (their cost was included as a noise factor) are as follows: (1) cold roller $800,000; (2) marker $50,000; and (3) robot $250,000. Average values, standard deviations, and S/N_{LTB} ratios were also coded for expediting the computation. Due to the size of the experiment Tables 6.5, 6.6, and 6.7 show only part of the complete experiment layout. Once the experiment was run, and the results were collected, it was possible to use this information for the succeeding analysis phase.

6.3.1 *Control factor array*

Taguchi separates variables into two types. Control factors are those variables that can be practically and economically controlled, such as dimensions and material parameters. Noise factors are variables that are difficult or expensive to control in practice, though they can be controlled in an experiment, such as the ambient temperature and the worker of a manufacturing plant. The objective is to determine the combination of control factor settings that will make the product have the maximum robustness to the expected variation in the noise factors. The measure of robustness is the signal-to-noise ratio, a log function of the desired output measurement. As shown in Table 6.5, to find the best levels of the control factors, the experiments, based on "orthogonal arrays," are balanced with respect to all control factors and yet are minimum in number.

An outstanding issue in robust parameter design is the choice of experimental plan. Most robust design experiments have used product arrays for the joint study of control and noise factors. In these designs, separate arrays are chosen for the control and the noise factors, and then each combination in the control factor array is paired with each combination in the noise factor array, producing a matrix of data. When the number of factors is large, a large experimental run size is needed for product arrays, making experiments costly and impractical. Recent statistical research on robust design experiments has favored single arrays, which include effects for both control factors and noise factors.

6.3.2 *Noise factor array*

Most robust design experiments have used product arrays for the joint study of design and noise factors. The analysis recommended by Taguchi involves two steps:

1. Summarize the data for each design factor combination
2. Study how the summary measure depends on the design factors

To test for dispersion effects of the design factors, Taguchi typically uses the logarithm of the coefficient of variation as a summary statistic.

Table 6.5 Partial View of the Control Factor Array Used in the Case Study

Run number	Associates (×1)	Delivery frequency Product A (×2)	Delivery frequency Product B (×3)	Delivery frequency Product C (×4)	Safety stock Product A (×5)	Safety stock Product B (×6)	Safety stock Product C (×7)	Number of machines (×8)	Number of setups (×9)
1	1	1	1	1	2	2	2	2	8
2	2	2	3	1	6	6	6	3	11
3	3	3	2	1	4	4	4	4	14
4	1	1	1	2	6	6	4	2	11
5	2	2	3	2	4	4	2	3	14
6	3	3	2	2	2	2	6	4	8
7	1	1	1	3	4	4	6	2	14
8	2	2	3	3	2	2	4	3	8
9	3	3	2	3	6	6	2	4	11
10	1	3	3	1	4	4	2	2	11
11	2	1	2	1	2	2	6	3	14
12	3	2	1	1	6	6	4	4	8
13	1	3	3	2	2	2	4	2	14
14	2	1	2	2	6	6	2	3	8
15	3	2	1	2	4	4	6	4	11
16	1	3	3	3	6	6	6	2	8
17	2	1	2	3	4	4	4	3	11
18	3	2	1	3	2	2	2	4	14
19	1	2	2	1	6	6	2	2	14

continued

Table 6.5 (continued) Partial View of the Control Factor Array Used in the Case Study

Run number	Associates (×1)	Delivery frequency Product A (×2)	Delivery frequency Product B (×3)	Delivery frequency Product C (×4)	Safety stock Product A (×5)	Safety stock Product B (×6)	Safety stock Product C (×7)	Number of machines (×8)	Number of setups (×9)
20	2	3	1	1	4	4	6	3	8
21	3	1	3	1	2	2	4	4	11
22	1	2	2	2	4	4	4	2	8
23	2	3	1	2	2	2	2	3	11
24	3	1	3	2	6	6	6	4	14
25	1	2	2	3	2	2	6	2	11
26	2	3	1	3	6	6	4	3	14
27	3	1	3	3	4	4	2	4	8
28	2	1	1	2	2	4	4	4	14
29	3	2	3	2	6	2	2	2	8
30	1	3	2	2	4	6	6	3	11
31	2	1	1	3	6	2	6	4	8
32	3	2	3	3	4	6	4	2	11
33	1	3	2	3	2	4	2	3	14
34	2	1	1	1	4	6	2	4	11

35	3	2	3	1	2	4	6	2	14
36	1	3	2	1	6	2	4	3	8
37	2	3	3	2	4	6	4	4	8
38	3	1	2	2	2	4	2	2	11
39	1	2	1	2	6	2	6	3	14
40	2	3	3	3	2	4	6	4	11
41	3	1	2	3	6	2	4	2	14
42	1	2	1	3	4	6	2	3	8
43	2	3	3	1	6	2	2	4	14
44	3	1	2	1	4	6	6	2	8
45	1	2	1	1	2	4	4	3	11
46	2	2	2	2	6	2	4	4	11
47	3	3	1	2	4	6	2	2	14
48	1	1	3	2	2	4	6	3	8
49	2	2	2	3	4	6	6	4	14
50	3	3	1	3	2	4	4	2	8

Table 6.6 Partial View of the Noise Factor Array and Set of Responses from the Case Study

Noise Factor Array							
LOW	LOW	HIGH	HIGH	HIGH	HIGH	HIGH	HIGH
HIGH	HIGH	LOW	LOW	LOW	HIGH	HIGH	HIGH
HIGH	HIGH	HIGH	HIGH	LOW	HIGH	LOW	LOW
LOW	HIGH	HIGH	LOW	HIGH	LOW	HIGH	LOW
HIGH	LOW	LOW	HIGH	HIGH	LOW	LOW	HIGH
HIGH	HIGH	LOW	HIGH	HIGH	LOW	HIGH	LOW
LOW	HIGH	HIGH	HIGH	LOW	LOW	LOW	HIGH
HIGH	LOW	HIGH	LOW	HIGH	HIGH	LOW	LOW
Response 5	**Response 6**	**Response 7**	**Response 8**	**Response 9**	**Response 10**	**Response 11**	**Response 12**
1675787	1700073	229967	235874	251372	–910227	–870396	–888516
2973289	3016566	2312869	2323332	2328624	2139059	2189535	2161560
2925659	2987927	2284365	2299381	2294459	2116580	2177694	2139862
1554637	1578927	30808	36711	52183	–1204135	–1164331	–1182455
2974727	3018002	2314309	2324773	2330074	2141193	2191679	2163705
2923040	2985308	2284075	2299091	2294169	2117617	2178731	2140898
1486786	1511095	–70618	–64734	–49394	–1350153	–1310481	–1328624
2972871	3016133	2314855	2325331	2330721	2142971	2193544	2165583
2937598	2999862	2296840	2311860	2306964	2131398	2192538	2154709
1595928	1620217	101846	107749	123221	–1098116	–1058312	–1076436
2942679	2985972	2275814	2286260	2291437	2090633	2140995	2113003
2914239	2976506	2263153	2278169	2273247	2084071	2145185	2107352
1520471	1544779	–14311	–8427	6913	–1266189	–1226517	–1244660
2950993	2994270	2284607	2295068	2300351	2101962	2152430	2124453
2911391	2973658	2263010	2278027	2273114	2085744	2146867	2109036
1591054	1615326	85974	91894	107490	–1125199	–1085272	–1103378
2953362	2996638	2288748	2299210	2304503	2107580	2158056	2130081
2912660	2974927	2265260	2280278	2275364	2088923	2150046	2112214
1562155	1586462	54371	60256	75605	–1164155	–1124474	–1142616
2972565	3015844	2318080	2328540	2333815	2150297	2200756	2172778
2903683	2965951	2250358	2265374	2260451	2066570	2127684	2089851
1666702	1690974	220798	226719	242315	–920939	–881011	–899117
2956783	3000078	2306692	2317135	2322295	2140326	2190671	2162676
2915700	2977965	2260430	2275449	2270544	2076832	2137964	2100134
1550088	1574379	25914	31816	47279	–1210541	–1170745	–1188870
2989292	3032553	2335016	2345493	2350891	2169533	2220115	2192155
2913357	2975621	2260869	2275889	2270993	2079581	2140721	2102892

Table 6.6 (continued) Partial View of the Noise Factor Array and Set of Responses from the Case Study

Noise Factors				
1. Raw material price	LOW	LOW	LOW	LOW
2. Interest rate	LOW	LOW	LOW	HIGH
3. Energy cost	LOW	LOW	HIGH	LOW
4. Tooling cost	LOW	LOW	HIGH	HIGH
5. Machine cost	LOW	LOW	HIGH	HIGH
6. Expected life	LOW	HIGH	LOW	LOW
7. Salvage value	LOW	HIGH	LOW	HIGH
8. Wage	LOW	HIGH	LOW	HIGH
Run number	**Response 1**	**Response 2**	**Response 3**	**Response 4**
1	2181964	2177412	2134237	1697781
2	3124707	3110939	3069889	2994304
3	3094463	3071480	3032565	2945686
4	2097492	2092941	2049800	1576605
5	3125887	3112120	3071058	2995750
6	3091339	3068355	3029441	2943067
7	2046841	2042290	1999319	1508622
8	3123666	3109898	3068723	2993982
9	3105539	3082555	3043607	2957650
10	2125205	2120653	2077512	1617896
11	3098371	3084603	3043700	2963579
12	3087427	3064443	3025528	2934265
13	2069735	2065184	2022213	1542306
14	3105847	3092079	3051040	2971999
15	3083897	3060913	3021988	2931426
16	2124816	2120264	2076964	1613145
17	3107664	3093896	3052846	2974377
18	3084811	3061827	3022901	2932695
19	2098530	2093978	2050996	1583999
20	3121573	3107806	3066778	2993562
21	3078759	3055775	3016861	2923709
22	2173600	2169049	2125749	1688793
23	3105116	3091349	3050468	2977665
24	3090723	3067739	3028802	2935744
25	2093540	2088988	2045859	1572047
26	3137571	3123804	3082617	3010413
27	3087510	3064526	3025578	2933409

Table 6.7 Columns of Average Profit and Standard Deviation from the Case Study

Run number	Average profit	Standard deviation	Run number	Average profit	Standard deviation
1	$801,277	$1,260,619	42	$2,666,028	$428,551
2	$2,645,389	$427,105	43	$2,644,594	$420,390
3	$2,614,177	$419,377	44	$747,868	$1,261,535
4	$626,599	$1,344,161	45	$2,648,005	$428,090
5	$2,646,940	$426,812	46	$2,622,502	$426,464
6	$2,612,928	$417,625	47	$534,337	$1,358,581
7	$535,079	$1,381,997	48	$2,649,304	$433,726
8	$2,646,523	$425,060	49	$2,636,320	$427,383
9	$2,626,760	$418,123	50	$672,513	$1,297,348
10	$688,114	$1,312,844	51	$2,657,169	$432,716
11	$2,609,754	$435,220	52	$2,617,513	$426,318
12	$2,596,132	$429,161	53	$657,630	$1,303,088
13	$584,291	$1,357,650	54	$2,645,316	$434,893
14	$2,618,758	$433,906	55	$2,594,077	$432,837
15	$2,594,923	$427,009	56	$2,662,421	$427,735
16	$676,090	$1,323,585	57	$744,733	$1,274,916
17	$2,622,247	$432,374	58	$2,593,915	$435,032
18	$2,596,826	$426,115	59	$2,627,453	$426,825
19	$644,592	$1,328,464	60	$760,501	$1,261,029
20	$2,648,533	$421,503	61	$2,593,957	$434,858
21	$2,583,752	$432,317	62	$2,652,175	$425,633
22	$791,969	$1,261,400	63	$579,907	$1,346,718
23	$2,635,105	$418,651	64	$2,636,629	$415,389
24	$2,594,835	$433,235	65	$2,646,200	$434,409
25	$621,646	$1,345,063	66	$662,619	$1,310,415
26	$2,665,788	$420,435	67	$2,627,414	$416,600
27	$2,594,246	$430,835	68	$2,627,892	$435,387
28	$2,597,972	$435,186	69	$610,490	$1,332,529
29	$697,218	$1,286,753	70	$2,641,708	$417,934
30	$2,675,508	$419,922	71	$2,627,376	$434,859
31	$2,614,158	$436,046	72	$765,798	$1,265,346
32	$542,152	$1,360,056	73	$2,621,133	$425,231
33	$2,679,267	$418,925	74	$2,667,003	$420,304
34	$2,607,348	$435,751	75	$483,491	$1,395,074
35	$572,839	$1,336,877	76	$2,611,906	$426,336
36	$2,669,284	$420,454	77	$2,653,516	$419,264
37	$2,651,063	$417,274	78	$739,549	$1,276,253
38	$665,962	$1,292,938	79	$2,616,960	$425,505

Table 6.7 (continued) Columns of Average Profit and Standard Deviation from the Case Study

Run number	Average profit	Standard deviation	Run number	Average profit	Standard deviation
39	$2,643,691	$426,504	80	$2,661,296	$421,823
40	$2,648,064	$417,416	81	$599,100	$1,343,340
41	$431,233	$1,409,339			

Noise factors are those system parameters that are difficult or costly to control and are presumed uncontrollable. Robust parameter design involves choosing optimal levels of the controllable factors in order to obtain a target or optimal response with minimal variation. As shown by Table 6.6, noise factors bring variability into the system, thus affecting the response.

For robust design, the objective is to properly choose the levels of control factors so that the process is robust or insensitive to the variation caused by noise factors. Robust parameter design methods are used to make systems more reliable and robust to incoming variations in environmental effects, manufacturing processes, and customer usage patterns. However, robust design can become expensive, time consuming, and resource intensive. Thus, research that makes robust design less resource intensive and requires a fewer number of experimental runs is of great value. Robust design methodology can be expressed as a multiresponse optimization problem.

6.3.3 *Average profit and standard deviation*

To implement robust design, Taguchi advocates the use of an "inner array" and "outer array" approach.

- The inner array consists of the orthogonal array that contains the control factor settings.
- The outer array consists of the orthogonal array that contains the noise factors and their settings that are under investigation.

The combination of the inner array and outer array constitutes what is called the "product array" or "complete parameter design layout." The product array is used to systematically test various combinations of the control factor settings over all combinations of noise factors after which the mean response and standard deviation may be approximated for each run, as shown in Table 6.7.

Between the mean (average) and standard deviation, it is typically easy to adjust the mean on target, but reducing the variance is difficult.

As shown by Table 6.7, the designer should minimize the variance first and then adjust the mean on target. Among the available control factors, most of them should be used to reduce variance. Only one or two control factors are adequate for adjusting the mean on target.

6.4 Analyzing a robust design

Regression analysis is typically carried out with the aid of a statistics software package. In this study, SAS was used to obtain the regression coefficients required for the full quadratic model in Equation 6.1. The regression analysis was performed by using the data contained in the control factor array and the column with the mean values of the responses across the noise factor array. The regression model is shown in Table 6.8. With help of the statistics software package, it was determined that the regression model obtained in this study had an adjusted R^2 statistic of 99.9%.

The analyses of the residual plots provided enough evidence of the normality and randomness of the residuals as well as the significance of the regression model. The regression model (in Table 6.8) was used to

Table 6.8 Regression Model Based on the Average Values of the Responses from the Case Study

Control factor	Regression coefficient	Control factor	Regression coefficient	Control factor	Regression coefficient
Constant	−8,443,162	X_1X_2	930	X_3X_8	12,867
X_1	−34,522	X_1X_3	5,119	X_3X_9	−3,224
X_2	−77,217	X_1X_4	−4,746	X_4X_5	1,679
X_3	50,815	X_1X_5	−2,324	X_4X_6	−7,169
X_4	−72,974	X_1X_6	−2,012	X_4X_7	−2,712
X_5	−19,080	X_1X_7	5,517	X_4X_8	23,247
X_6	35,629	X_1X_8	10,702	X_4X_9	−127
X_7	−22,118	X_1X_9	451	X_5X_6	−4,238
X_8	6,714,718	X_2X_3	645	X_5X_7	−279
X_9	−42,246	X_2X_4	24,644	X_5X_8	7,809
X_1^2	−8,276	X_2X_5	5,478	X_5X_9	96
X_2^2	−4,799	X_2X_6	−1,057	X_6X_7	1,900
X_3^2	−9,395	X_2X_7	3,400	X_6X_8	3,676
X_4^2	−355	X_2X_8	−972	X_6X_9	38
X_5^2	208	X_2X_9	2,847	X_7X_8	3,966
X_6^2	−1,421	X_3X_4	−6,434	X_7X_9	−825
X_7^2	354	X_3X_5	106	X_8X_9	12,087
X_8^2	−1,001,918	X_3X_6	−2,230		
X_9^2	−62	X_3X_7	−2,127		

predict the average profit value of all the possible combinations of control factors included in the study. The full factorial enumeration and the prediction of the average profit values were carried out in a spreadsheet. A total of 3^9 = 19,683 combinations or designs were evaluated. The average profit values were used as the criterion to sort the designs in decreasing order and from this ranking the 15 most profitable designs were selected. These designs and their predicted profit values are shown in Table 6.9.

Regarding the Taguchi methods, the objective of this type of analysis is to find the combination of control factors that has the lowest variation across the combinations of noise factors (a robust design) and the largest average response. Taguchi methods make use of the S/N_{LTB} ratio defined by Equation 5.2. In this study, the set of responses need to be transformed into positive values to adequately compute the S/N_{LTB} ratios. The transformed set of responses, the control factor array, and the S/N_{LTB} ratios were then used to generate the control factor effect plots from which a design was elicited.

The combination of all the points at which the S/N_{LTB} ratio is maximized in each of the control factor effect plots is considered to be the optimal design according to the Taguchi methods. The design generated by use of the Taguchi methods analysis was identified in the full factorial enumeration from the regression analysis in order to have a basis of comparison with the rest of the alternative designs. The combination number 5134 matched the Taguchi methods generated design. The average predicted profit value for combination number 5134 was $2,646,625, and compared to the designs selected from the regression analysis, it occupied the 16th place (see Table 6.9).

6.5 *Evaluation based on the confirmation run*

The fourth phase of the method is evaluation, its objective is to select the best design from the alternatives generated in the analysis phase. For this purpose, it is necessary to perform confirmation runs, interact with the decision maker to include practical issues, and finally, to let the decision maker decide which manufacturing cell design is best for the intended application. Once the optimum choice has been made, it is tested by performing a confirmation run.

This run is used to "validate" the model as well as confirm the improvements in the process. During a confirmation run, the following is performed:

- Run two additional cases with starting and optimum control factor levels
- Compare the predicted and actual results

Table 6.9 16 Most Profitable Designs from the Analysis Phase

Combination number	Number of associates (×1)	Delivery frequency Product A (×2)	Delivery frequency Product B (×3)	Delivery frequency Product C (×4)	Safety stock Product A (×5)	Safety stock Product B (×6)	Safety stock Product C (×7)	Number of machines (×8)	Number of setups (×9)	Predicted average profit
5755	1	3	2	3	6	2	2	3	8	$2,777,210
5760	1	3	2	3	6	2	2	4	14	$2,776,015
6484	1	3	3	3	6	2	2	3	8	$2,773,533
787	1	1	2	1	2	6	2	3	8	$2,773,274
1516	1	1	3	1	2	6	2	3	8	$2,771,831
796	1	1	2	1	2	6	4	3	8	$2,769,468
805	1	1	2	1	2	6	6	3	8	$2,768,494
7366	2	1	2	1	2	6	6	3	8	$2,767,662
5031	1	3	1	3	6	2	2	4	14	$2,767,379
6489	1	3	3	3	6	2	2	4	14	$2,765,861
1525	1	1	3	1	2	6	4	3	8	$2,763,771
8095	2	1	3	1	2	6	6	3	8	$2,762,830
5026	1	3	1	3	6	2	2	3	8	$2,762,097
76	1	1	1	1	2	6	6	3	8	$2,759,655
5756	1	3	2	3	6	2	2	3	11	$2,759,216
5134 (Taguchi Methods)	1	3	2	1	2	4	2	3	8	$2,646,625

For example, an electroless nickel deposition process on ceramic substrates was optimized using a Taguchi-based robust design method. The plating parameter settings were obtained in only 27 experiments. The savings in experiments is significant when compared to a full factorial design of 486 runs. The yield of a completely metallized substrate was substantially increased. Under optimal conditions, a 100% yield was obtained in the confirmation run of 80 ceramic samples. For the case study, the optimum parameter setting obtained through experimentation resulted in robust broad banded design, which was confirmed by conducting the confirmation run. Sixteen confirmation runs were performed, one for each of the resulting designs. Profit values were obtained across the original noise factor array to obtain a confirmed average profit and a confirmed standard deviation for each design. The results are shown in Table 6.10.

Figure 6.2 shows a plot of the predicted and the confirmed average profit values of the 16 resulting designs and Figure 6.3 shows a plot of their standard deviations. By examination of these plots, it can be concluded that the first eight designs (6484, 5755, 5756, 5026, 6489, 5760, 5031, and 5134) are more robust than the rest of the designs.

The combination number 5134 of the Taguchi methods design ended up being the eighth best option in terms of confirmed average profit (from an initial 16th place) and fourth best in standard deviation.

The evaluation of the designs obtained from the confirmation runs was then presented to the decision maker, who expanded the criteria of the selection to include the following two practical considerations:

1. Staffing constraints in the operation
2. Raw material supply issues

Regarding staff, the new manufacturing cell represents a challenge to the typical operations: people working in the cell must undergo a series of training courses and a change in mind-set to adapt to working in the new environment. This situation makes it difficult to rely on having only one operator in the cell in the short term; it is necessary to start with two associates and maybe reevaluate the more profitable designs with one associate in the future.

The number of deliveries of raw material should be minimized, since truck-receiving labor is a limitation in the plant. These two issues were taken into account along with the previous analyses for a final decision.

6.6 Decision making based on evaluation

The decision maker was finally able to select a specific design after assessing the potential benefits and risks of all the alternatives. The chosen design was combination number 8095 generated by regression analysis.

Table 6.10 Ranking of the 16 Designs after the Confirmation Runs

Combination number	Predicted average profit	Confirmed average profit	Confirmed standard deviation	Original ranking based on predicted average profit	New ranking based on confirmed average profit	New ranking based on confirmed standard deviation
6484	$2,773,533	$2,696,380	$418,849	3	1	2
5755	$2,777,210	$2,687,462	$419,531	1	2	6
5756	$2,759,216	$2,687,365	$419,025	15	3	3
5026	$2,762,097	$2,682,951	$422,222	13	4	7
6489	$2,765,861	$2,679,684	$417,577	10	5	1
5760	$2,776,015	$2,674,482	$419,380	2	6	5
5031	$2,767,379	$2,669,234	$422,482	9	7	8
5134	$2,646,625	$2,664,002	$419,379	16	8	4
8095	$2,762,830	$2,651,935	$433,713	12	9	10
1516	$2,771,831	$2,650,990	$434,370	5	10	11
1525	$2,763,771	$2,647,527	$434,439	11	11	12
805	$2,768,494	$2,639,097	$436,108	7	12	16
787	$2,773,274	$2,638,882	$435,565	4	13	14
796	$2,769,468	$2,638,056	$435,928	6	14	15
7366	$2,767,662	$2,625,549	$435,183	8	15	13
76	$2,759,655	$2,611,906	$426,336	14	16	9

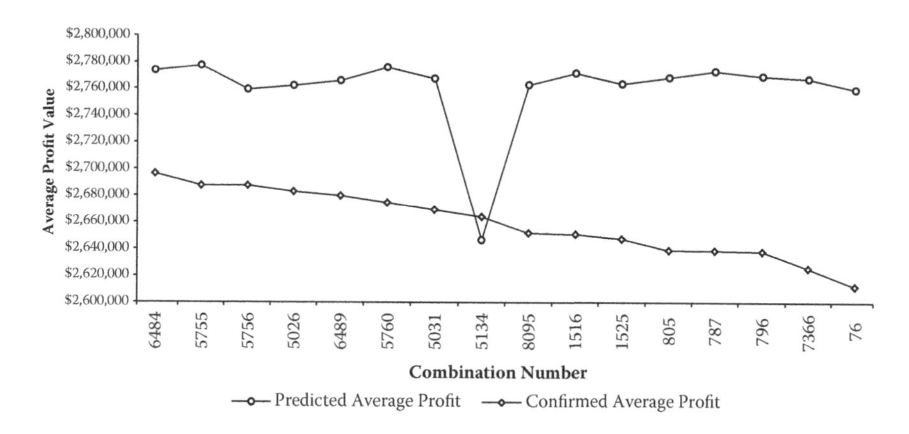

Figure 6.2 The predicted and confirmed average profit of the 16 best designs.

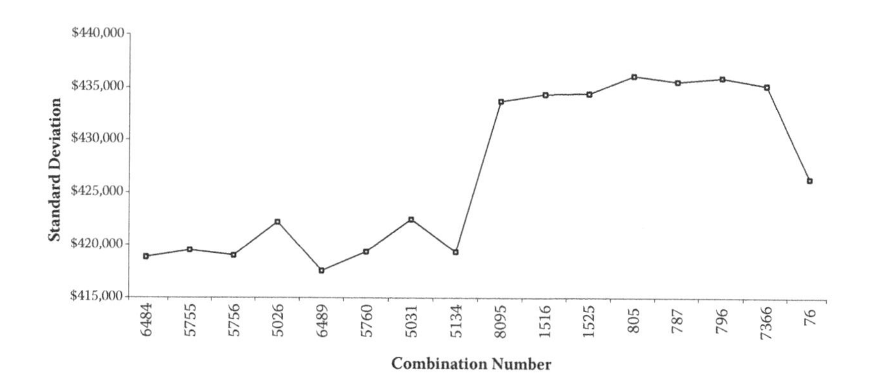

Figure 6.3 The confirmed standard deviations of the 16 best designs.

The final design establishes in terms of staffing levels that two associates must be staffed in the manufacturing cell. In terms of inventory policies, it establishes that the raw material for product A must be replenished once per week and a safety stock equivalent to two shifts of production (approximately 960 blanks) must be maintained on hand; that the raw material for product B must be replenished three times a week and a safety stock equivalent to six shifts of production (approximately 2880 blanks) must be maintained on hand; and that the raw material for product C must be replenished once per week and a safety stock equivalent to six shifts of production (approximately 2880 blanks) must be maintained on hand. It also establishes that three CNC lathes should be installed for the second operation, and, finally, it is recommended to keep the number of setups at eight for every 6 weeks.

The expected average profit under the stated assumptions is $2,651,935 for every 6 weeks (approximately $23,071,834 per annum). The cost of the decision to choose this design instead of the optimal design is $44,445 for every 6 weeks (approximately $386,671 per annum).

In this case study, a new method was successfully used to partially design a manufacturing cell at a local manufacturing plant aiming for the profit maximization of the cell. The methodology was used to generate several designs with high estimated average profit values and low sensitivity to changes in selected noncontrollable economic factors. The generation of these designs allowed the design maker to choose the best design according to the profit value, robustness, and other practical considerations.

The study included an optimal setting of number of associates; raw material inventory policy (frequency, order size, and safety stock); number of machines for a specific operation; and number of setups per planning period. The methodology integrated simulation techniques, design of experiments, regression analysis, Taguchi methods, and a new profit model to facilitate the generation and evaluation of different designs.

The final evaluation of the resulting designs suggested that Taguchi methods is a viable and simple way to generate cell designs with high profit values and low variability to noise factors. It is also evident that regression analysis can be used to generate a wide span of combinations with trade-offs between them that was clearly reflected in the profitability of each design. In conclusion, the results of this study suggest that the design of a manufacturing cell can be carried out by using potential profitability of the system as the central objective.

6.7 How to incorporate networked security into a cellular manufacturing system

The architecture of large electrical supervisory control and data acquisition (SCADA) systems is moving toward distributed processing. An energy management system (EMS) tends to be large and requires many network interfaces. Servers handle

- Automatic generation control (AGC)
- Load forecasting, energy accounting, and outage management
- Sequence of events (SOE)

The architecture of smaller electrical systems tends to be simpler and contain fewer network connections such as

- Servers
- Operator interfaces

- Remote terminal units (RTUs)/programmable logic controllers (PLCs)
- Intelligent electronic devices (IEDs)

The SCADA network is formed when the clients, servers, and RTUs/controllers communicate using some type of protocol and network connections. Modern SCADA systems typically have multiple network connections. Network connections for SCADA systems can be subdivided into two general categories: "outside" and "inside." Network connections may consist of modem, leased lines, optic microwave, radio, or direct network connections.

Inside network connections may communicate with many types of devices, including RTUs, IEDs, other SCADAs, modems, PLCs, and GPS clocks. Inside network protocols include Distributed Network Protocol (DNP) 3.0 (Serial and Ethernet), Modbus, Transmission Control Protocol (TCP)/Internet Protocol (IP), legacy protocols, and proprietary protocols. Outside network connections may communicate with multiple systems including other EMS and regional transmission organizations (RTOs). Outside network protocols include Inter-control Center Communications Protocol (ICCP), File Transfer Protocol (FTP), Distributed Network Protocol (DNP) 3.0 (Serial and Ethernet), TCP/IP, Modbus, and IEC 61850. Note that TCP/IP is not a substation protocol but rather an underlying data transfer method comprised of a suite of protocols for application on Internet/intranet. Substation protocol selection depends on a variety of factors, such as the selected SCADA system, user preference, and available protocols in other SCADA systems and devices. SCADA systems have the availability to communicate with more than one protocol. A SCADA system can communicate with one device with one protocol and another device with a different protocol. The most popular protocol for use outside SCADA network connections is ICCP. ICCP is based on an international standard (IEC60870-6 TASE.2) that enables integration of real-time data between the control centers of different utilities over Wide Area Networks (WANs). Leased lines are another popular network connection between control centers. ICCP-TASE.2 is being used worldwide within the electric and gas utility industries to provide exchange of real-time data between control centers, substations, power plants, EMS, SCADA, and metering equipment. Applications range from small regional power dispatching networks to large multinational transmission systems.

6.8 One-piece flow (OPF): A critical systems way of thinking

Lean involves a fundamental paradigm shift from conventional batch-and-queue mass production to product-aligned one-piece flow (OPF) pull production. Whereas batch-and-queue involves the mass production of large lots of products in advance based on potential or predicted customer demands: an OPF system rearranges production activities in a way that processing steps of different types are conducted immediately adjacent to each other in a continuous flow.

This shift requires highly controlled processes operated in a well-maintained, ordered, and clean environment that incorporates principles of employee-involved, system-wide, continual improvement.

While most of these methods are interrelated and can occur concurrently, most organizations begin by implementing Lean techniques in a particular production area or at a pilot facility, and then expand the use of the methods over time. Companies typically tailor these methods to address their own unique needs and circumstances. In doing so, they may develop their own terminology for the various methods.

Bibliography

Agarwal, A., and Sarkis, J. 1998. A review and analysis of comparative performance studies on functional and cellular manufacturing layouts. *Computers and Industrial Engineering*, 34, 77–98.

Ang, C. L., and Willey, P. C. T. 1984. A comparative study of the performance of pure and hybrid group technology manufacturing systems using computer simulation techniques. *International Journal of Production Research*, 22(2), 193–233.

Arinez, J. F., and Cochran, D. S. 1999. Application of a production system design framework to equipment design. *32nd CIRP International Seminar on Manufacturing Systems*.

Askin, R. G., and Subramanian, S. P. 1987. A cost-based heuristic for group technology configuration. *International Journal of Production Research*, 25, 101–113.

Askin, R. G., Selim, H. M., and Vakharia, A. J. 1987. A methodology for designing flexible cellular manufacturing systems. *IIE Transactions*, 29, 599–610.

Assad, A. A., Kramer, S. B., and Kaku, B. K. 2003. Comparing functional and cellular layouts: A simulation study based on standardization. *International Journal of Production Research*, 41(8), 1639–1663.

Atwater, J. B., and Chakravorty, S. 1995. Using the theory of constraints to guide the implementation of quality improvement projects in manufacturing operations. *International Journal of Production Research*, 25, 1737–1760.

Black, J. T. 1991. *The Design of the Factory with a Future*. New York: McGraw-Hill.

Buchwald, S. 1994. Throughput accounting: A recipe for success. *1994 Conference Proceedings of American Production and Inventory Control Society*, pp. 635–637.

Byrne, D. M., and Taguchi, S. 1987. Taguchi approach to parameter design. *Quality Progress*, 20, 19–26.

Cary, D. F. 1993. What the big boys really think about you: How cycle time reduction shows up in the financial analysis. *1993 Conference Proceedings of American Production and Inventory Control Society*, pp. 704–707.

Cochran, D. 1999. The production system design and deployment framework. SAE Technical paper 1999-01-1644, *SAE IAM-99 Conference*.

Connor, W. S., and Zelen, M. 1969. *Fractional Factorial Experimental Designs for Factors at Three Levels*. Washington, D.C.: National Bureau of Standards.

Dean, B. R., Kaye, M., and Hand, S. 1997. A cost-based strategy for assessing improvements in manufacturing processes. *International Journal of Production Research*, 35, 955–968.

Evren, R. 1987. Interactive compromise programming. *Journal of the Operational Research Society*, 38, 163–172.

Flynn, B. B., and Jacobs, F. R. 1986. A simulation comparison of group technology with traditional job shop manufacturing. *International Journal of Production Research*, 24(5), 1171–1192.

Fowlkes, W. Y., and Creveling, C. M. 1995. *Engineering Methods for Robust Product Design*. New York: Addison-Wesley.

Fry, T. D. 1995. Japanese manufacturing performance criteria. *International Journal of Production Research*, 33, 933–954.

Gershon, M. 1984. The role of weights and scales in the application of multiobjective decision making. *European Journal of Operational Research*, 15, 244–250.

Goicoechea, A., Hansen, D. R., and Duckstein, L. 1982. *Multiobjective Decision Analysis with Engineering and Business Applications*. New York: Wiley & Sons.

Goldratt, E. M., and Cox, J. 1984. *The Goal: A Process of Ongoing Improvement*. Great Barrington, MA: North River Press.

Harris, C. R. 1997. Modeling the impact of design, tactical, and operational factors on manufacturing system performance. *International Journal of Production Research*, 35, 479–499.

Keeney, R. L., and Raiffa, H. 1976. *Decisions with Multiple Objectives*. New York: Wiley & Sons.

Kelton, D. W., Sadowski, R. P., and Sadowski, D. A. 1998. *Simulation with Arena*. New York: McGraw-Hill.

Lee, S., and Jung, H. J. 1989. A multiobjective production planning model in a flexible manufacturing environment. *International Journal of Production Research*, 27, 1981–1992.

Lewis, H. S., Sweigart, J. R., and Markland, R. E. 1996. An interactive decision framework for multiple objective production planning. *International Journal of Production Research*, 34, 3145–3164.

Liggett, H. R., and Trevino, J. 1992. The application of multi-attribute techniques in the development of performance measurement and evaluation models for cellular manufacturing. *Flexible Automation and Information Management, 1992*, pp. 711–722.

Monden, Y. 1993. *Toyota Production System*. Norcross, GA: Industrial Engineering and Management Press.

Montgomery, D. C. 1997. *Design and Analysis of Experiments*. New York: John Wiley & Sons.

Morris, J. S., and Tersine, R. J. 1990. A simulation analysis of factors influencing the attractiveness of group technology cellular layouts. *Management Science*, 36(12), 1567–1578.

Mosca, R., Giribone, P., and Drago, A. 1992. A simulation approach to FMS dimensioning including economic evaluations. *Flexible Automation and Information Management, 1992*, pp. 771–781.

Needy, K. L., Billo, R. E., and Colosimo Warner, R. L. 1998. A cost model for the evaluation of alternative cellular manufacturing configurations. *Computers and Industrial Engineering*, 34, 119–134.

Park, H., and Lee, T. 1995. Design of a manufacturing cell in consideration of multiple objective performance measures. In *Planning, Design, and Analysis of Cellular Manufacturing Systems*, edited by A. K. Kamrani, H. R. Parsaei, and D. H. Liles. New York: Elsevier.

Pillai, V. M., and Subbarao, K. 2008. Robust cellular manufacturing system design for dynamic part population using a genetic algorithm. *International Journal of Production Research*, 46(18), 5191–5210.

Raffish, N. 1994. Activity-based costing—Part II. *1994 Conference Proceedings of American Production and Inventory Control Society*, pp. 631–633.

Romero, C., Tamiz, M., and Jones, D. F. 1998. Goal programming, compromise programming, and reference point method formulations: Linkages and utility interpretations. *Journal of the Operational Research Society*, 49, 986–991.

Sarfaraz, A. R., and Emamizadeh, B. 1993. Product costing for the concurrent engineering environment. *1993 Conference Proceedings of American Production and Inventory Control Society*, pp. 688–690.

Seifoddini, H., and Djassemi, M. 1997. Determination of flexibility range for cellular manufacturing systems under product mix variations. *International Journal of Production Research*, 35(2), 3349–3366.

Selim, H. M., Askin, R. G., and Vakharia, A. J. 1998. Cell formation in group technology: Review, evaluation and directions for future research. *Computers and Industrial Engineering*, 14, 3–20.

Shi, Y., and Yu, P. L. 1989. Goal setting and compromise solutions. In *Multiple Criteria Decision Making and Risk Analysis Using Microcomputers*, edited by B. Karpak and S. Zionts. Berlin, Germany: Springer-Verlag.

Singh, N. 1993. Design of cellular manufacturing systems: An invited review. *European Journal of Operational Research*, 69, 284–291.

Stevens, M. E. 1993. Activity-based costing/management (ABC/ABM): The strategic weapon of choice in the global war competition. *1993 Conference Proceedings of American Production and Inventory Control Society*, pp. 699–703.

Suh, N. P. 1990. *Principles of Design*. New York: Oxford University Press.

Suresh, N. 1990. Towards an integrated evaluation of flexible automation investments. *International Journal of Production Research*, 28, 1657–1672.

Wang, J. X. 2002. *What Every Engineer Should Know about Decision Making under Uncertainty*. Boca Raton, FL: CRC Press.

Wang, J. X. 2005. *Engineering Robust Designs with Six Sigma*. Upper Saddle River, NJ: Prentice Hall.

Wang, J. X. 2010. *Lean Manufacturing: Business Bottom-Line Based*. Boca Raton, FL: CRC Press.

Wang, J. X. 2012. *Green Electronics Manufacturing: Creating Environmental Sensible Products*. Boca Raton, FL: CRC Press.

Wang, J. X., and Roush, M. L. 2000. *What Every Engineer Should Know about Risk Engineering and Management*. Boca Raton, FL: CRC Press.

Wemmerlov, U., and Hyer, N. L. 1989. Cellular manufacturing in the U.S. industry: A survey of users. *International Journal of Production Research*, 27, 1511–1530.

Wemmerlov, U., and Johnson, D. J. 1997. Cellular manufacturing at 46 user plants: Implementation experiences and performance improvements. *International Journal of Production Research*, 35, 29–49.

Wicks, E. M., and Reasor, R. J. 1999. Designing cellular manufacturing systems with dynamic part populations. *IIE Transactions*, 31, 11–20.

Wizdo, A. 1993. A methodology for relevant costing and profitability. *1993 Conference Proceedings of American Production and Inventory Control Society*, pp. 694–698.

Womack, J. P., Roos, D., and Jones, D. T. 1990. *The Machine That Changed the World*. New York: Rawson Associates.

Yu, P. L. 1985. *Multiple-Criteria Decision Making: Concepts, Techniques, and Extensions*. New York: Plenum Press.

Zionts, S. 1989. Multiple criteria mathematical programming: An updated overview and several approaches. In *Multiple Criteria Decision Making and Risk Analysis Using Microcomputers*, edited by B. Karpak and S. Zionts. Berlin, Germany: Springer-Verlag.

chapter seven

Networked cellular manufacturing systems

The automated control of manufacturing devices on a common communications network is becoming a necessity for cellular manufacturing. The Open Systems Interconnection (OSI) protocols represent a major effort by a group of companies toward meeting this need and apparently is becoming a standard for factory-floor communications. Cellular manufacturing is in transition, moving from the traditional approach that relies on independent fixed-automation machines to a more flexible and systematic approach that can adapt and move rapidly with consumers' changing demands.

7.1 Open systems interconnection

The use of automated manufacturing cells that utilize programmable computer controlled devices is now a recognized method for implementing this new approach to manufacturing. Manufacturers also realize the importance of inter-machine and intracell communication to facilitate automated and flexible manufacturing systems.

Modern factories generally use programmable devices from a variety of vendors, each with a different proprietary protocol. To enable these different machines to "talk" to each other, custom interfaces must be designed to translate one vendor's protocol into another's. Networking several different machines within a factory can be a time-consuming and expensive process. The OSI reference model is the best known and most widely used guide for visualizing networking environments. Manufacturers adhere to the Open Systems Interconnection reference model when they design network products. It provides a description of how network hardware and software work together in a layered fashion to make communications possible. The model also helps to troubleshoot problems by providing a frame of reference that describes how components are supposed to function. The OSI model describes how data is sent and received over a network.

The OSI model is a suggested standard for communication that was developed by the International Organization for Standardization (ISO). The OSI reference model describes how data is sent and received over a network. This model breaks down data transmission over a series of seven layers. Each layer has a responsibility to perform specific tasks concerning sending and receiving data. All the layers are needed for a message to reach its destination. The OSI model provides software developers with a standard for developing communications software. The OSI model also offers the standard for communications so that different manufacturers' computers can be used on the same network.

For example, robots are widely used in the automotive industry and related industries for tasks such as material handling, spray painting and finishing, electronics and mechanical assembly, and spot and arc welding. Robots are also used in unmanned manufacturing cells, where one robot may tend to several computer-controlled machine tools. Communications between each machine tool and the robot is an essential component of manufacturing cells, and the degree to which this communication is accomplished often determines the success and payback potential of the cell. Communication between cells further enhances the flexibility and utility of each cell. OSI is being promoted as the communication standard for the factory floor. With OSI as the standard, machine tools from different vendors could share production scheduling data, part information, and materials resource planning (MRP) data, and cell controllers could download machine tool programs, monitor performance, and coordinate cell activities.

An unmanned manufacturing cell is a group of machine tools and associated material-handling equipment managed by a local supervisory computer or cell controller. The cell manufactures part components that belong to one or more part families. Being unmanned, the cell operates at least semiautonomously; that is, some operations within the cell, such as cell initialization and error recovery, may require human intervention. A robot often performs material handling tasks between machines within the cell, including part transfer and inspection, tool replacement, and product assembly. The cell may involve metalworking operations, such as turning, drilling, boring, and milling; finishing operations, such as deburring and painting; and assembly or other operations.

The cell operates untended for extended intervals under the cell controller's supervision. The cell controller directs the flow of parts through the cell, monitors and controls the operation of the individual machine tools, and manages communications between machines in the cell. When this cell is implemented in a factory, the cell controller will also be the "spokesman" for the group of machine tools to the other cells on the factory floor and to the factory-level supervisory computer. The use of OSI enables different makes of programmable devices and factory-floor computers to communicate with each other.

7.2 Seven layers of the OSI model

OSI consists of seven layers: the physical layer, data link layer, network layer, transport layer, session layer, presentation layer, and application layer.

Software and hardware can then be separately developed for each layer. However, they must work together to successfully transport a message. Each layer performs a specific function, but all of the layers have one function in common: communicating with the layers above and below them in the model.

The application layer is the software that the end user interacts with. This application is like Firefox, Outlook, or Internet Explorer. This layer provides service to applications outside the OSI model. It performs several functions, including establishing the availability of the communication partner, synchronizing the sending and receiving of applications, establishing agreement on error recovery and data integrity, and determining if sufficient resources exist for the communication to occur.

The presentation layer is concerned with the presentation of data. This layer defines the format that the data uses as it is transmitted. It formats the data for the user so that it is readable and the message can be understood. This layer may also compress data for easier transmission or encrypt data for security purposes.

The session layer is responsible for allowing ongoing communication between two parties across the network. It handles the setup of the session, data exchanges, and the end of the session. This layer is responsible for flow control, or defining the rules for communication between two computers. Flow control will prevent too much data from being sent to the receiving computer at one time so it does not become overloaded.

The transport layer, also known as the end-to-end layer, deals with the transmission of data between networks. The transport layer ensures that error-free data is given to the user. This layer generates the address for the receiving computer and adds it to the data so that it is sent to the correct destination. It sets priorities for messages and error recovery procedures in the event that an error on the network occurs.

The network layer splits up long messages into smaller bits of data, often referred to as packets. The network layer chooses the route that data will take and addresses the data for delivery. It adds a destination address and routing information to enable the packet to travel between nodes on the network.

The data link layer moves information from one computer or network to another computer or network. It performs three specific functions: controls the physical layer and decides when to transmit messages, formats messages indicating where they start and end, and detects and corrects errors that occur during transmission.

Whereas the upper six layers are concerned with software, the physical layer is concerned with hardware. The physical layer provides the physical connection between the computer and network. The physical components may include servers, clients, and circuits.

For the OSI model, each layer performs a specific function, but all the layers have one function in common: communicating with the layers above and below them. The purpose of the OSI model is to give software developers a standard for developing communications software so that different manufacturers' computers can be used on the same network.

7.3 Error detection and correction for networked cellular manufacturing systems

7.3.1 Messages in a networked cellular manufacturing system

For networked cellular manufacturing systems, error detection and correction are techniques that enable reliable delivery of digital data over unreliable communication channels. Many communication channels are subject to channel noise, and thus errors may be introduced during transmission from the source to a receiver.

- Messages are a sequence of bits, 0s and 1s.
- Bit corruption—channel changes the value of some bits.

$$0 \rightarrow 1 \qquad 1 \rightarrow 0$$

- Corruption may be random for each bit (thermal noise, for example).
- Corruption also may occur in bursts.
- Corruption burst is the smallest subsequence containing all corrupted bits.

Example 7.1

```
message sent:        01101110011011
message received:    01111110111001
corrupted bits:         X    X   X
corruption burst:       |--------|
```

7.3.2 Error detection and correction

Error detection techniques allow detecting such errors, while error correction enables reconstruction of the original data. The general definitions of the terms are as follows:

- *Error detection* is the detection of errors caused by noise or other impairments during transmission from the transmitter to the receiver. Error detection detects whether a message is corrupted (not whether individual bits are corrupted). The ability to perform error detection is usually measured in two ways:

1. We say that a protocol performs x-bit detection, iff, for any message, and for any y number of corrupted bits in the message, where $y \leq x$, the protocol will detect the corruption.
2. We say that a protocol performs x-burst detection, iff, for any message, and for any corruption burst of size y in the message, where $y \leq x$, the protocol will detect the corruption.

- *Error correction* is the detection of errors and reconstruction of the original, error-free data. Error correction corrects a corrupted message (restores its original value).
  1. We say a protocol performs x-bit correction, iff, in any message, if any y number of bits are corrupted, where $y \leq x$, the protocol can restore the original contents of the message.
  2. Note that error bursts are not considered.

  Being accomplished by adding redundant check bits, error correction may generally be realized in two different ways:
- *Automatic repeat request* (ARQ; sometimes also referred to as a backward error correction)—This is an error control technique whereby an error detection scheme is combined with requests for retransmission of erroneous data. Every block of data received is checked using the error detection code, and if the check fails, retransmission of the data is requested. This may be done repeatedly, until the data can be verified.
- *Forward error correction* (FEC)—The sender encodes the data using an error-correcting code (ECC) prior to transmission. The additional information (redundancy) added by the code is used by the receiver to recover the original data. In general, the reconstructed data is what is deemed the "most likely" original data.

ARQ and FEC may be combined such that minor errors are corrected without retransmission and major errors are corrected via a request for retransmission: this is called a hybrid automatic repeat request (HARQ).

7.3.3 Error detection and correction schemes

Several schemes exist to achieve error detection. The general idea is to add some *redundancy* (i.e., some extra data) to a message, which enables detection of any errors in the delivered message. Most such error detection schemes are systematic: The transmitter sends the original data bits and attaches a fixed number of *check bits*, which are derived from the data bits by some deterministic algorithm. The receiver applies the same algorithm to the received data bits and compares its output to the received check bits; if the values do not match, an error has occurred at some point during the transmission. In a system that uses a "non-systematic" code, such as some raptor codes, the original message is transformed into an encoded message that has at least as many bits as the original message.

- Code word: Sequence M, |M| = m, data bits followed by a sequence R, |R| = r, check bits (usually m >> r)

M	R
m data bits	r check bits

- The R check bits are a function of the M data bits, that is, there is a standard function f, known by both the sender and receiver.
- A code word is valid, iff, R = F(M).

A hash function F is a transformation that takes a variable-size input m and returns a fixed-size string, which is called the hash value h (that is, h = F(M)). Hash functions with just this property have a variety of general computational uses, but when employed in cryptography the hash functions are usually chosen to have some additional properties. In general, any hash function may be used to compute the redundancy. However, some functions are of particularly widespread use because of either their simplicity or their suitability for detecting certain kinds of errors (e.g., the cyclic redundancy check's performance in detecting burst errors).

Error-correcting codes can provide a suitable alternative to hash functions when a strict guarantee on the minimum number of errors to be detected is desired. Repetition codes, described next, are special cases of error-correcting codes: although rather inefficient, they find applications for both error correction and detection due to their simplicity.

7.3.4 Repetition codes

A repetition code is a coding scheme that repeats the bits across a channel to achieve error-free communication. Given a stream of data to be transmitted, the data is divided into blocks of bits. Each block is transmitted a predetermined number of times.

Example 7.2

For example, to send the bit pattern "1011", the 4-bit block can be repeated three times, thus producing "1011 1011 1011". However, if this 12-bit pattern was received as "1010 1011 1011"—where the first block is unlike the other two—it can be determined that an error has occurred.

Repetition codes are not very efficient and can be susceptible to problems if the error occurs in exactly the same place for each group (e.g., "1010

1010 1010" in the previous example would be detected as correct). The advantage of repetition codes is that they are extremely simple and are in fact used in some transmissions of numbers stations.

7.3.5 Parity bits

A parity bit is a bit that is added to ensure that the number of set bits (i.e., bits with the value 1) in a group of bits is even or odd. The simplest case is by adding a parity bit. Suppose we have a 3-bit word (so the bit strings define points in a cube). If we add a fourth bit, we can decree that any time we want to switch a bit in the original 3-bit string, we also have to switch the parity bit. A parity bit can only detect an odd number of errors (i.e., one, three, five, etc., bits that are incorrect). The procedure is as follows:

- Sender computes R = f(M), and sends M;R (; = concatenation)
- Channel transforms (corrupts) M;R into M′;R′, and receiver receives M′;R′
- Receiver computes X = f(M′)
- If X = R′, the receiver accepts the msg., if X ≠ R′, the receiver rejects the message
- Note, X ≠ R′ ⇒ corruption, but ¬ (X = R′ ⇒ no corruption), since the channel may turn a valid code word into another valid code word

There are two variants of parity bits: *even parity bit* and *odd parity bit.* When using even parity, the parity bit is set to 1 if the number of ones in a given set of bits (not including the parity bit) is odd, making the entire set of bits (including the parity bit) even. When using odd parity, the parity bit is set to 1 if the number of ones in a given set of bits (not including the parity bit) is even, making the entire set of bits (including the parity bit) odd. In other words, an even parity bit will be set if the number of set bits plus one is even, and an odd parity bit will be set if the number of set bits plus one is odd.

There is a limitation to parity schemes. A parity bit is only guaranteed to detect an odd number of bit errors. If an even number of bits (i.e., two, four, six, etc.) are flipped, the parity bit will appear to be correct even though the data is erroneous. Extensions and variations on the parity bit mechanism are horizontal redundancy checks, vertical redundancy checks, and "double," "dual," or "diagonal" parity (used in RAID-DP).

7.3.6 Checksum concept

One approach of error checking is to append the sum value of all message bytes to the end of the message. A checksum is a fixed-size datum computed from an arbitrary block of digital data for the purpose of detecting accidental errors that may have been introduced during its transmission or storage. The integrity of the data can be checked at any later time by recalculating the checksum and comparing it with the stored one. If the checksums match, the data was likely not accidentally altered. This sum can identify the message and changes in its contents. On the other hand, if there is more than one change; one that adds up a value ($0 \geq 1$) and one that subtracts one ($1 \geq 0$) in a way that the sum remains the same, so it cannot be used to detect errors. The same can happen if the checksum is changed with the same value as the message.

A checksum of a message is a modular arithmetic sum of message code words of a fixed word length (e.g., byte values). The sum may be negated by means of a one's-complement prior to transmission to detect errors resulting in all-zero messages.

- Code is a collection of legal code words
- Assume for simplicity that you always send m data bits
- m data bits $\Rightarrow 2^m$ legal code words
- That is, $2^{m+r} - 2^m$ = number of illegal code words
- For example, if m = 8 and r = 2, 256 valid code words, 1024 – 256 invalid code words
- The hope is that if corruption occurs, the channel turns a valid code word into an invalid code word (not another valid code word), since there are more invalid code words than valid ones
- The larger the number of illegal code words (i.e., larger r) the better

Checksum schemes include parity bits, check digits, and longitudinal redundancy checks. Some checksum schemes, such as the Luhn algorithm and the Verhoeff algorithm, are designed specifically to detect errors commonly introduced by humans when writing or remembering identification numbers.

7.3.7 *Hamming distance*

Error code detection and correction are widely used in data storage devices (disks and RAM, for example) and in data communication systems. Even though the probability of a single-bit error occurring (due to electromagnetic interference, dynamic RAM memory fading, for example) can be quite small, the fact that large amounts of data (collections of bits) are stored and transmitted makes it probable that errors are occurring during transmission or storage.

For example, if the probability that a single bit is wrong ("corrupted") is represented by p, then the probability that a single bit is not wrong is $1 - p$. The probability of a sequence of n uncorrupted bits is thus $(1 - p)^n$. If

$p = 10^{-6}$ ("one in a million") this probability, of an uncorrupted message of 1,000,000 bits, is only $(0.999999)^{1,000,000} = 0.37$ (approximately). In the case of long bit sequences, errors are thus likely and need to be dealt with.

A simple way to detect an error is to add to m data bits (carrying the message) a parity bit. This bit is chosen to have a value (0 or 1) in such a way that the corresponding number of data bits that are '1' is even ("even parity") or odd ("odd parity") in the string of m + 1 bits. For example, if one chooses even parity and the 3 words of 4 data bits are to be transmitted:

4 data bits	parity bit
0 1 1 1	1
0 0 1 1	0
1 0 0 1	0

This allows one to detect only errors that consist of an odd number of wrong bits but does not allow one to determine which bit(s) is (are) wrong. One could construct a code with more than one parity bit, for example, by forming a matrix where the last column consists of parity bits and a row is added that consists of parity bits. For the previous three words (parity bits are in bold):

0 1 1 1 **1**
0 0 1 1 **0**
1 0 0 1 **0**
1 1 0 1 1

In this way, a single error in a row or column can be located. There is, however, a price to pay. In the example case, there are 8 parity bits for a total of 12 data bits. Obviously, as the word size increases the extra overhead becomes smaller. For small word sizes, the overhead of the parity bits may lead to less efficiency than simply detecting the error and retransmitting.

A general way to detect and correct errors is by using the Hamming distance. It employs the Hamming distance between bit configurations. The Hamming distance between two bit strings of length n is defined as the number of bits that differ in the two strings. It can be defined compactly as

$$\text{Hamming distance } H(s_1, s_2) = |s_1[n-1] - s_2[n-1]| + |s_1[n-2] - s_2[n-2]| + |s_1[1] - s_2[1]| + |s_1[0] - s_2[0]|$$

Here, I have used the usual enumeration of the digits in the bit strings: the LSB (least significant bit) has and index (subscript) of 0 and the MSB (most significant bit) is 1. If all bits have the same value, the Hamming distance is 0; if they are all different, the Hamming distance is n.

The 2^n different bit strings that can be formed from n bits can be represented in a binary tree in such a way that the Hamming distance has an immediate geometrical interpretation. Hamming distance of two equal-length code words is the number of different bits in two code words.

Example 7.3

```
c1 = 1000100
c2 = 1011001
xor = 0011101
Hamm(c1, c2) = 4
```

In summary, the Hamming distance of a code is the minimum hamming distance of any pair of equal-length code words in the code.

The best starting point for understanding ECC codes is to consider bit strings as addresses in a binary hypercube. A hypercube is a generalization of a cube to various dimensions; we are probably most familiar with the notion of a four-dimensional hypercube. Figure 7.1 shows images of binary hypercubes for several different dimensionalities.

In Figure 7.1, each of the binary hypercubes was created by copying the one to the left twice, and connecting the corresponding vertices.

We can assign each vertex in a hypercube a location in a coordinate space determined by the dimensionality of the hypercube.

- A zero-dimensional hypercube requires no coordinates to know where you are.
- A one-dimensional hypercube can use one bit to tell whether you are at the bottom or the top of the line segment.
- A two-dimensional hypercube can use two bits: first bit is left versus right, second is inherited from the line.
- A three-dimensional hypercube can use a bit to tell front square from back, and inherit two bits from the square.
- A four-dimensional hypercube can use a bit to tell left cube from right, and inherits three bits from the cube.

This can continue through as many dimensions as you want. The Hamming distance between two bit strings is the number of bits you have

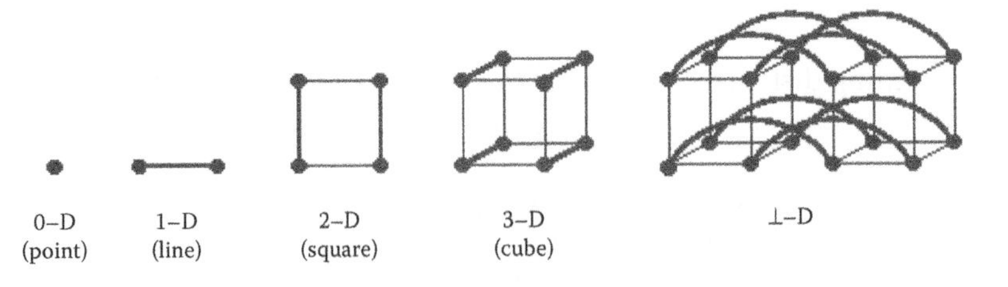

Figure 7.1 A binary hypercube.

to change to convert one to the other; this is the same as the number of edges you have to traverse in a binary hypercube to get from one of the vertices to the other. The basic idea of an error-correcting code is to use extra bits to increase the dimensionality of the hypercube, and make sure that the Hamming distance between any two valid points is greater than one.

- If the Hamming distance between valid strings is only one, a single-bit error results in another valid string. This means we cannot detect an error.
- If it is two, then changing one bit results in an invalid string and can be detected as an error. Unfortunately, changing just one more bit can result in another valid string, which means we cannot know which bit was wrong, so we can detect an error but not correct it.
- If the Hamming distance between valid strings is three, then changing one bit leaves us only one bit away from the original error, but two bits away from any other valid string. This means if we have a one-bit error, we can figure out which bit is the error; but if we have a two-bit error, it looks like one bit from the other direction. So we can have single bit correction, but that is all.
- Finally, if the Hamming distance is four, then we can correct a single-bit error and detect a double-bit error. This is frequently referred to as a SECDED (single-error correct, double-error detect) scheme.

To perform x-bit detection, Hamm(code) $\geq x + 1$. Why? Assume Hamm(code) $\geq x + 1$, then all valid code words are separated by at least $x + 1$ bit changes.

```
       ≥ x+1              ≥ x+1
   |<--------------->|<--------------->|
   c2---|-----------c1-----------|----c3
        e                  e'
```

If the sender sends c1, and the channel corrupts x bits or less (resulting in words e or e′), you end up with an invalid word, because other valid words (e.g., c2 and c3) are at least $x + 1$ bits away from c1.

Example 7.4

Byte parity bit ("even" parity bit of a sequence of bits is the XOR of these bits) $m = 8$ data bits and $r = 1$ parity bit

```
              byte      parity (even)
code word            10111010 1
single bit error 11111010 1 = illegal code word
double bit error 11011010 1 = another valid code word
```

The Hamming distance of a byte parity bit code = 2, and it performs a one-bit detection. How can the Hamming distance be increased between valid strings? A common technique that is used to increase the Hamming distance is a cyclic redundancy check (CRC).

7.4 Error correction

7.4.1 Automatic repeat request

An automatic repeat request (ARQ) is an error control method for data transmission that uses error-detection codes, acknowledgment or negative acknowledgment messages, and timeouts to achieve reliable data transmission. An *acknowledgment* is a message sent by the receiver to indicate that it has received a data frame correctly.

Usually, when the transmitter does not receive the acknowledgment before the timeout occurs (i.e., within a reasonable amount of time after sending the data frame), it retransmits the frame until it is either received correctly or the error persists beyond a predetermined number of retransmissions.

Three types of ARQ protocols include: Stop-and-Wait ARQ, Go-Back-N ARQ, and Selective Repeat ARQ.

ARQ is appropriate if the communication channel has varying or unknown capacity, such as is the case on the Internet. However, ARQ requires the availability of a backchannel, results in possibly increased latency due to retransmissions, and requires the maintenance of buffers and timers for retransmissions, which in the case of network congestion can put a strain on the server and overall network capacity.

7.4.2 Error-correcting codes

Any error-correcting code can be used for error detection. A code with a *minimum* Hamming distance, *d*, can detect up to *d*-1 errors in a code word. Using error-correcting codes for error detection can be suitable if a strict limit on the minimum number of errors to be detected is desired.

Example 7.5

To perform an x-bit correction, you replace the received invalid code word by the *closest* legal code word (in terms of Hamming distance).

```
E.g. valid code words 00000 00000
                                    00000 11111
                                    11111 00000
                                    11111 11111

code Hamm dist = 5
Sender sends                00000 00000
```

```
Receiver receives          00000 00011
Receiver corrects to       00000 00000
              (closest legal code word to received code word)
```

Codes with minimum Hamming distance $d = 2$ are degenerate cases of error-correcting codes and can be used to detect single errors. The parity bit is an example of a single-error-detecting code. The Berger code is an early example of a unidirectional error(-correcting) code that can detect any number of errors on an asymmetric channel, provided that only transitions of cleared bits to set bits or set bits to cleared bits can occur.

7.4.3 Forward error correction

An error-correcting code (ECC) or forward error correction (FEC) code is a system of adding redundant data, or *parity data,* to a message, such that it can be recovered by a receiver even when a number of errors (up to the capability of the code being used) were introduced, either during the process of transmission or on storage. Since the receiver does not have to ask the sender for retransmission of the data, a backchannel is not required in forward error correction, and it is therefore suitable for simplex communication such as broadcasting. Error-correcting codes are frequently used in lower-layer communication, as well as for reliable storage in media such as CDs, DVDs, and dynamic RAM. Error-correcting codes are usually distinguished between convolutional codes and block codes:

- *Convolutional codes* are processed on a bit-by-bit basis. They are particularly suitable for implementation in hardware, and the Viterbi decoder allows optimal decoding.
- *Block codes* are processed on a block-by-block basis. Early examples of block codes are repetition codes, Hamming codes, and multidimensional parity-check codes. They were followed by a number of efficient codes, of which Reed-Solomon codes are the most notable ones due to their widespread use these days. Turbo codes and low-density parity-check codes (LDPC) are relatively new constructions that can provide almost optimal efficiency.

Shannon's theorem is an important theorem in forward error correction, and describes the maximum information rate at which reliable communication is possible over a channel that has a certain error probability or signal-to-noise ratio (SNR). This strict upper limit is expressed in terms of the channel capacity. More specifically, the theorem says that there exist codes such that with increasing encoding length the probability of error on a discrete memoryless channel can be made arbitrarily small, provided that the code rate is smaller than the channel capacity. The code rate is defined as the fraction k/n of k source symbols and n encoded symbols.

The actual maximum code rate allowed depends on the error-correcting code used and may be lower. This is because Shannon's proof was only existential in nature and did not show how to construct codes that are both optimal, and have efficient encoding and decoding algorithms.

7.4.3.1 *Hamming distance for correction*

To perform an x-bit correction, the Hamming distance of the code needs to be at least 2x + 1 (i.e., the Hamming distance of the code is ≥ 2x + 1). Why? If x bits corrupted, there is only *one valid* code word within a distance of x bits of the invalid code word. Assume Hamm(code) ≥ 2x + 1, and the sender sends the code word c1.

```
|<-------2x+1---->|<--------2x+1------->|
c2---------------c1--------|-----------c3
                ≤ x ≥ x+1
            |<------>|<----------->|
                  e
```

e = c1 corrupted by x bits
c2 and c3 are at least x + 1 away from e
c1 is only x or less away from e (thus correct to c1)

Assuming only x bits or less can get corrupted, the closest word to e is c1, and the receiver correctly corrects e to c1.

Example 7.6

In Example 7.5, we can always correct ≤2 bit errors, since

Hamm(code) = 5.

E.g., sender sends	00000 11111
Receiver receives	00000 **00**111
Receiver corrects to	00000 11111

This is assuming ≤2 bits are corrupted.

If sender sends	00000 11111
And the receiver receives	00000 **000**11 (3 bit errors)
The receiver *incorrectly* corrects the code word to	00000 00000

7.4.4 *Hybrid schemes*

A hybrid ARQ is a combination of an ARQ and forward error correction. There are two basic approaches:

- Messages are always transmitted with FEC parity data (and error-detection redundancy). A receiver decodes a message using the parity information, and requests retransmission using ARQ only if the parity data was not sufficient for successful decoding (identified through a failed integrity check).
- Messages are transmitted without parity data (only with error-detection information). If a receiver detects an error, it requests FEC information from the transmitter using ARQ and uses it to reconstruct the original message.

The latter approach is particularly attractive on the binary erasure channel when using a rateless erasure code.

7.4.5 Applications of error correction methods

Applications that require low latency (such as telephone conversations) cannot use ARQ; they must use FEC. By the time an ARQ system discovers an error and retransmits it, the resent data will arrive too late to be any good.

Applications where the transmitter immediately forgets the information as soon as it is sent (such as most television cameras) cannot use ARQ; they must use FEC because when an error occurs, the original data is no longer available. (This is also why FEC is used in data storage systems such as RAID and distributed data store.)

- Applications that use ARQ must have a return channel. Applications that have no return channel cannot use ARQ.
- Applications that require extremely low error rates (such as digital money transfers) must use ARQ.

7.5 Cyclic redundancy check (CRC) for an OSI data link

All communications systems try to make sure that the transmitted messages reach the destination without any problem. So they intend to implement different algorithms in order to satisfy this requirement. One of the options is to encode the message in a way that enables the receiver to check the corrupted messages; an example of such coding is the cyclic redundancy check (CRC). For the OSI model, packets are placed into frames at this layer. The CRC is added at this layer. If the CRC fails at the receiving computer, this layer will request retransmission. Mac addresses are resolved at this layer.

The data links provide transparent network services to the network layer so the network layer can ignore the underlying physical network topology. It is responsible for reassembling bits, taken off the wire by the physical layer, to frames, and makes sure they are in the correct order and requests retransmission of frames in case an error occurs. It provides error checking by adding a CRC to the frame, and flow control. Examples of devices that operate on this layer include switches, bridges, wireless access points (WAPs), and network interface cards (NICs).

Data storage devices must also prevent any corruption of its data. The best choice for this problem is to use data redundancy, which costs a lot. So instead, data can be encoded using redundant codes in order to detect the corrupted data; and CRC is one of the most common codes that is used in such cases.

7.5.1 What is CRC?

CRC stands for cyclic redundancy check, which means that it is based on a cyclic algorithm that generates redundant information. CRC is an error-checking code that is widely used in data communications systems and other serial data transmission systems. CRC is based on polynomial manipulations using modulo arithmetic. Some of the common CRC standards are CRC-8, CRC-12, CRC-16, CRC-32, and CRC-CCIT.

For cellular manufacturing, the potentially unreliable physical link between two machines is converted into an apparently very reliable link by including redundant information in each transmitted frame. Depending on the nature of the link and the data, one can include just enough redundancy to make it possible to detect errors and then arrange for the retransmission of damaged frames. The CRC is a widely used parity-bit-based error detection scheme in serial data transmission applications. This code is based on polynomial arithmetic.

- To each message, we will append n additional bits, where n is a constant.
- Let m be the message size. Note that m is not constant; the technique works for messages of any size.
- We assume we have a constant G, which is a string of n + 1 bits.
- These n + 1 bits can be viewed as the coefficients of an *n*th degree polynomial.

For example, if G = 1011 (n = 3), then G can be thought of as the polynomial.

$$\mathbf{1} \rightarrow x^3 + \mathbf{0} \rightarrow x^2 + \mathbf{1} \rightarrow x + \mathbf{1} \rightarrow x^0$$

$$x^3 + x + 1$$

The CRC performs a mathematical calculation on a block of data and returns information (number) about the contents and organization of that data. So the resultant number uniquely identifies that block of data. This unique number can be used to check the validity of data or to compare two blocks. So this approach is used in many communication and computer systems to ensure the validity of the transmitted or stored data. In general CRC codes are able to detect:

- All single- and double-bit errors
- All odd numbers of errors
- All burst errors less than or equal to the degree of the polynomial used
- Most burst errors greater than the degree of the polynomial used

7.5.2 CRC idea

The main idea of CRC is to treat the message as binary numbers and divide it by a fixed binary number. The remainder from this division is considered the checksum. The recipient of the message performs the same division and compares the remainder with the checksum (transmitted remainder). The CRC is a simple binary division and subtraction. The only difference is that these operations are done on modulo arithmetic based on mod 2. For example, the addition and subtraction are replaced with the XOR operation that does the sum and subtraction without carry.

7.5.2.1 Polynomial concept

The CRC algorithm uses the term polynomial to perform all of its calculations. This polynomial is the same concept as the traditional arithmetic polynomial. The divisor, dividend, quotient, and remainder that are represented by numbers are represented as polynomials with binary coefficients.

For example, the number 23 (10111b) can be represented in the polynomial form as:

$$1{*}x^4 + 0{*}x^3 + 1{*}x^2 + 1{*}x^1 + 1{*}x^0$$

or

$$x^4 + x^2 + x^1 + x^0$$

Note the binary representation of the number (10111).

This representation simplifies the traditional arithmetic operations (addition, multiplication, etc.) that are all done on normal algebraic polynomials. If we can assume that X is 2, then the operations are simplified more and some terms can be canceled. For example, the term $3{*}x^3$ is represented as 24 in a normal number representation and $24 = 16 + 8$, which is $x^4 + x^3$ in polynomial representation.

Example 7.7

Here is a simple polynomial, $2x^2 - 3x + 7$. It is a function of some variable x, which depends only on powers of x. The *degree* of a polynomial is equal to the highest power of x in it; here it is 2 because of the x^2 term. A polynomial is fully specified by listing its coefficients, in this case (2, –3, 7). Notice that to define a degree-d polynomial you have to specify d + 1 coefficients. It is easy to multiply polynomials. For instance,

$$(2x^2 - 3x + 7) \times (x + 2) = 2x^3 + 4x^2 - 3x^2 - 6x + 7x + 14$$

$$= 2x^3 + x^2 + x + 14$$

Conversely, it is also possible to divide polynomials. For instance, the above equation can be rewritten as a division:

$$(2x^3 + x^2 + x + 14)/(x + 2) = 2x^2 - 3x + 7$$

Just like in integer arithmetic, one polynomial does not have to be divisible by another. But you can always divide out the "whole" part and be left with the remainder. For instance, $x^2 - 2x$ is not divisible by x + 1, but you can calculate the quotient to be x – 3 and the remainder to be 3:

$$(x^2 - 2x) = (x + 1) \times (x - 3) + 3$$

In fact, you can use a version of long division to perform such calculations.

7.5.2.2 Generator polynomial

In order to do the CRC calculation a divisor must be selected, which can be any one. This divisor is called the *generator polynomial*. Even though, some polynomials became standard for many applications, polynomial selection is beyond the scope of this chapter.

One of the most used terms in CRC is the width of the polynomial. This width is represented by the order of the highest power in the polynomial. The width of the polynomial in the previous example is 4, which has 5 bits in its binary representation.

Since CRC is used to detect errors, a suitable generator polynomial must be selected for each application. This is because each polynomial has different error detection capabilities. CRC algorithms are commonly called after the generator polynomial width, for example CRC-16 uses a generator polynomial width of 15 and a 16-bit register, and CRC-32 uses a polynomial width of 31 and a 32-bit register.

Most of us are familiar with polynomials whose coefficients are real numbers. In general, however, you can define polynomials with coefficients taken from arbitrary sets. One such set (in fact a *field*) consists of the numbers 0 and 1 with arithmetic defined modulo 2. It means that you perform arithmetic as usual, but if you get something greater than 1 you keep only its remainder after division by 2. In particular, if you get 2, you keep 0. Here's the addition table:

$0 + 0 = 0$
$0 + 1 = 1 + 0 = 1$
$1 + 1 = 0$ (because 2 has a remainder of 0 after dividing by 2)

The multiplication table is equally simple:

$0 \times 0 = 0$
$0 \times 1 = 1 \times 0 = 0$
$1 \times 1 = 1$

Moreover, subtraction is also well defined (in fact the subtraction table is identical to the addition table) and so is division (except for division by zero). What is nice, from the point of view of computer programming, is that both addition and subtraction modulo 2 are equivalent to bitwise *exclusive OR* (XOR).

Now imagine a polynomial whose coefficients are zeros and ones, with the rule that all arithmetic on these coefficients is performed modulo 2. You can add, subtract, multiply, and divide such polynomials (they form a *ring*). For instance, let's do some easy multiplication:

$$(1x^2 + 0x + 1) \times (1x + 1) = 1x^3 + 1x^2 + 0x^2 + 0x + 1x + 1$$
$$= 1x^3 + 1x^2 + 1x + 1$$

Let's now simplify our notation by representing a polynomial as a series of coefficients. For instance, $1x^2 + 0x + 1$ has coefficients (1, 0, 1), $1x + 1$ (1, 1), and $1x^3 + 1x^2 + 1x + 1$ (1, 1, 1, 1).

A polynomial with coefficients modulo 2 can be represented as a series of bits. Conversely, any series of bits can be looked upon as a polynomial. In particular, any binary message, which is nothing but a series of bits, is equivalent to a polynomial.

7.5.2.3 Mathematical digression: Modulo-2 arithmetic

Taking a bitwise exclusive-OR in place of performing an addition is an example of "modulo-2 arithmetic," which is one form of "polynomial arithmetic." I have seen one author call it "CRC arithmetic."

Modulo-2 arithmetic is an arithmetic scheme; like most of the oddities that mathematicians like to study it seems completely useless to a non-mathematician at first glance but turns out to have some very practical applications. In this case, the practical application is in developing CRC checks. The basic idea of modulo-2 arithmetic is just that we are working in binary, but we do not have a carry in addition or a borrow in subtraction. This means:

- Addition and subtraction become the same operation: just a bitwise exclusive OR. Because of this, the total ordering we expect of integers is replaced by a partial ordering: one number is greater than another if and only if its leftmost 1 is farther left than the other's. This will have an impact on division, in a moment.
- Multiplication is just like multiplication in ordinary arithmetic, except that the adds are performed using exclusive ORs instead of additions.
- Division is like long division in ordinary arithmetic, except for two differences: the subtractions are replaced by exclusive ORs, and you can subtract any time the leftmost bits line up correctly (since, by the partial ordering described earlier, they are regarded as equal in this case).

It will probably help to show examples of modulo-2 multiplication and division.

7.5.2.4 Multiplication

```
   1101
   0110
   ----
   0000
  11010
 110100
0000000
-------
0101110
```

7.5.2.5 Division

```
         1101
     --------
0110)0101110
     0110
     ----
      0111
      0110
      ----
       0011
       0000
       ----
        0110
        0110
        ----
        0000
```

Notice that the first subtraction is possible in modulo-2 arithmetic, whereas it would not be possible in normal arithmetic. When we perform a modulo-2 addition on two numbers we get an answer of 0 or 1. In this case, we are performing the arithmetic on each coefficient of the polynomial modulo-2.

- Let M be the original data message (with m bits).
- M is just a string of bits, and can be viewed as a polynomial.
- Note that $M \rightarrow X^n$ = M concatenated with n zeros.

If M = 1001001 and n = 4, then $M \rightarrow X^n$ = 1001001**0000**.

- Let R be the remainder of $(M \rightarrow X^n)/G$.

Then, $(M \rightarrow X^n - R)/G$ should give a remainder of zero (simple arithmetic).

7.5.2.6 Procedure for CRC

Take a binary message and convert it to a polynomial, then divide it by another predefined polynomial called the *key*. The remainder from this division is the CRC. Now transmit both the message and the CRC. The recipient of the transmission does the same operation (divides the message by the same key) and compares his CRC with yours. If they differ, the message must have been mangled. If, on the other hand, they are equal, the odds are pretty good that the message went through uncorrupted. Most localized corruptions (burst of errors) will be caught using this scheme.

Not all keys are equally good. The longer the key, the better the error checking. On the other hand, the calculations with long keys can get pretty involved. Ethernet packets use a 32-bit CRC corresponding to a degree-31 remainder (remember, you need d + 1 coefficients for a degree-d polynomial). Since the degree of the remainder is always less than the degree of the divisor, the Ethernet key must be a polynomial of degree 32. A polynomial of degree 32 has 33 coefficients requiring a 33-bit number to store it. However, since we know that the highest coefficient (in front of x^{32}) is 1, we do not have to store it. The key used by the Ethernet is 0x04c11db7. It corresponds to the polynomial:

$$x^{32} + x^{26} + \ldots + x^2 + x + 1$$

There is one more trick used in packaging CRCs. First, calculate the CRC for a message to which you have appended 32 zero bits. Suppose that the message had N bits, thus corresponding to degree N – 1 polynomial. After appending 32 bits, it will correspond to a degree N + 31 polynomial. The top-level bit that was multiplying x^{N-1} will be now multiplying x^{N+31} and so on. In all, this operation is equivalent to multiplying the message polynomial by x^{32}. If we denote the original message polynomial by M(x), the key polynomial by K and the CRC by R(x) (remainder) we have:

$$M * x^{32} = Q(x) * K(x) + R(x)$$

Now add the CRC to the augmented message and send it away. When the recipient calculates the CRC for this sum, and there was no transmission error, he will get zero. That is because

$$M * x^{32} + R(x) = Q(x) * K(x) \text{ (no remainder!)}$$

Remember, in arithmetic modulo-2, addition and subtraction are the same. We will use this property of the CRC to test our implementation of the algorithm.

- The data D is multiplied by (X^n) and divided by the generator polynomial (G), the quotient (Q) is discarded, and the remainder (R) is considered the checksum.
- On the other side, the data stream (D) is multiplied again by X^n and the checksum (CRC) R is added to it (normally it comes with the stream) and the whole result is divided by G again.
- The result now should be zero for valid data.

This operation can be described by the following equation:

$$(X^n * D) + R = (Q * G) + 0$$

7.5.3 *How CRC works*

The CRC algorithm requires the division of the message polynomial by the key polynomial. The straightforward implementation follows the idea of long division, except that it is much simpler. The coefficients of our polynomials are ones and zeros. We start with the leftmost coefficient (leftmost bit of the message). If it is zero, we move to the next coefficient. If it is one, we subtract the divisor; except that subtraction modulo-2 is equivalent to exclusive OR, so it is very simple.

Let's do a simple example, dividing a message 100110 by the key 101. Remember that the corresponding polynomials are $x^5 + x^2 + x$ and $x^2 + 1$. Since the degree of the key is 2, we start by appending two zeros to our message. The calculation of dividing a message 100110 by the key 101 is summarized in Table 7.1.

We do not bother calculating the quotient; all we need is the remainder (the CRC), which in this case is 01. The original message with the CRC attached reads 10011001. As shown by the calculation in Table 7.1, it is easy to convince it that it is divisible by the key, 101, with no remainder.

Table 7.1 Dividing a Message 100110 by the Key 101

```
10011000/101
101
____
 111
 101
 ____
  100
  101
  ____
   100
   101
   ____
    01
```

7.5.3.1 Transmitter calculation

The transmitter can append zeros to the end of the message (LSB), perform the division, and find the remainder and append it to the original message.

- The sender computes R (i.e., the remainder of $(M^*X^n)/G$), and attaches R at the end of M, and sends it over the channel. That is, the sender sends M;R, where ; denotes concatenation.
- The channel transforms M;R into M′;R′.
- The receiver receives M′;R′ and performs $(M'^*X^n - R')/G$.
- Let z be the remainder of $(M'^*X^n - R')/G$.

If z is zero, then it assumes the message is not corrupted (although it may be). If z is not zero, then for sure the message is corrupted.

7.5.3.2 Receiver calculation

The message receiver can do one of the following:

- Separate the message and checksum. Calculate the checksum for the message (after appending the zeros) and compare the two checksums.
- Checksum the whole message including the CRC (without appending the zeros) and check if the new CRC comes out as zero.

Modulo-2 addition and subtraction are nothing more than bitwise exclusive OR (XOR).

Example 7.8

```
 1010      1010
+0110     -0110
-----     -----
 1100      1100
```

That is, + and – are both the same as XOR.

7.5.4 CRC implementation

CRC has two main implementation techniques:

- Serial implementation
- Parallel implementation

7.5.4.1 Serial implementation

This approach is a direct mapping for the CRC algorithm. In fact, this approach does not use standard microprocessor divide instruction because (1) we need XOR-based division (no carry in addition or subtraction) and (2) the dividend (the message) can be very large and behind the processor support. This approach is relatively low speed and consumes very small resources. This implementation is summarized in Table 7.2.

This approach can be implemented directly using linear-feedback shift registers (LFSR) where the division is performed by left shifting and subtraction by XOR.

- Let n = 4, thus |G| = n + 1 = 5, |R| = n = 4 (leading zeros of R included).
- Sender computes R = remainder of $(M^*X^n)/G$.

Table 7.2 Serial Implementation of CRC

```
To perform the division, do the following:
 Load the register with zero bits.
 Augment the message by appending W zero bits to the end
   of it.
 While (more message bits)
  Begin
  Shift the register left by one bit, reading the next bit
   of the augmented message into register bit position 0.
  If (a 1 bit popped out of the register during step 3)
   Register = Register XOR Poly.
  End
The register now contains the remainder.
```

- Sender sends M;R, receiver computes remainder of $(M*X^n - R)/G$. Note: $(M*X^n - R) = M;R$, because – equals XOR and $|R| = 4$, that is,

$$M*X^4 = M;0000, (M*X^4 - R) = (M;0000 - R) = M;R$$

- Thus, the receiver computes $(M*X^n - R)/G = (M;R)/G$.

This is convenient, because the division steps can be performed immediately as each bit of M;R is received.

7.5.4.2 Parallel implementation

In the real world, the serial approach (bit-by-bit) calculation is not acceptable for many applications that require high performance or if the smallest processing word size is more than a bit. For this reason, parallel implementation is needed.

Parallel implementation can be done by entering the data word to the CRC and performing the serial shift and XORing the bits in parallel. It simply calculates all XOR combinations as if a single bit shifting is done on the data and CRC register. Let us now analyze the power of this approach.

- Let E be the *error mask* caused by the channel, that is, a bit string indicating which bits got corrupted.
- Let F = M;R (the bit frame sent by the sender).
- The receiver receives F + E, which equals F XOR E (recall modulo-2).
- If there were no corrupted bits, $E = 0$. Otherwise, $E \neq 0$.

7.6 Single-minute exchange of die

Single-minute exchange of die (SMED) is a system for dramatically reducing the time it takes to complete equipment changeovers. The essence of the SMED system is to convert as many changeover steps as possible to "external" (performed while the equipment is running) and to simplify and streamline the remaining steps. The name single-minute exchange of die comes from the goal of reducing changeover times to the "single" digits (i.e., less than 10 minutes). A successful SMED program has the following benefits:

- Lower manufacturing cost (faster changeovers mean less equipment downtime)
- Smaller lot sizes (faster changeovers enable more frequent product changes)
- Improved responsiveness to customer demand (smaller lot sizes enable more flexible scheduling)
- Lower inventory levels (smaller lot sizes result in lower inventory levels)
- Smoother startups (standardized changeover processes improve consistency and quality)

SMED enables an organization to quickly convert a machine or process to produce a different product type. A single cell and set of tools can therefore produce a variety of products without the time-consuming equipment changeover and setup time associated with large batch-and-queue processes, enabling the organization to quickly respond to changes in customer demand.

Bibliography

Black, J. T. 1991. *The Design of the Factory with a Future*. New York: McGraw-Hill.

Black, J. T. 2000. Lean manufacturing implementation. In *Innovations in Competitive Manufacturing*, edited by Paul M. Swamidass, pp. 177–186. Boston: Kluwer Academic.

Brandon, J. 1996. *Cellular Manufacturing: Integrating Technology and Management*. Somerset, England: Research Studies Press LTD.

Burbidge, J. L. 1978. *The Principles of Production Control*. England: MacDonald and Evans.

Feld, W. M. 2001. *Lean Manufacturing: Tools, Techniques, and How to Use Them*. Boca Raton, FL: St. Lucie Press; Alexandria, VA: APICS.

Flanders, R. E. 1925. Design manufacture and production control of a standard machine. *Transactions of ASME*, 46, 691–738.

Hyer, N., and Wemmerlov, U. 2002. *Reorganizing the Factory: Competing through Cellular Manufacturing*. Portland, OR: Productivity Press.

Irani, S. 1999. *Handbook of Cellular Manufacturing Systems*. New York: John Wiley & Sons.

Mitrofanov, S. P. 1966. *The Scientific Principles of Group Technology*. Boston: National Lending Library for Science and Technology.

Peterson, W. W., and Brown, D. T. 1961. Cyclic codes for error detection. In *Proceedings of the IRE*, pp. 228–235.

Singh, N., and Rajamani, D. 1996. *Cellular Manufacturing Systems Design, Planning and Control*. London: Chapman & Hall.

Swamdimass, P. M., and Darlow, Neil R. 2000. Manufacturing strategy. In *Innovations in Competitive Manufacturing*, edited by P. M. Swamidass, pp. 17–24. Boston: Kluwer Academic.

Tanenbaum, A. S. 1988. *Computer Networks*, 2nd ed. Englewood Cliffs, NJ: Prentice Hall.

Wang, J. X. 2002. *What Every Engineer Should Know about Decision Making under Uncertainty*. Boca Raton, FL: CRC Press.

Wang, J. X. 2005. *Engineering Robust Designs with Six Sigma*. Upper Saddle River, NJ: Prentice Hall.

Wang, J. X. 2010. *Lean Manufacturing: Business Bottom-Line Based*. Boca Raton, FL: CRC Press.

Wang, J. X. 2012. *Green Electronics Manufacturing: Creating Environmental Sensible Products*. Boca Raton, FL: CRC Press.

Wang, J. X., and Roush, M. L. 2000. *What Every Engineer Should Know about Risk Engineering and Management*. Boca Raton, FL: CRC Press.

Womack, James P., Jones, Daniel T., and Roos, Daniel. 2001. *The Machine That Changed the World*. New York: Harper Perennial.

chapter eight

Risk engineering of a networked cellular manufacturing system

Networked cellular manufacturing systems (NCMSs) are typically found in industrial plants where remote monitoring and control of systems is necessary because of hazardous or inaccessible equipment locations. Such facilities might include water treatment plants, wastewater treatment plants, chemical engineering facilities, oil refineries, nuclear facilities, and other such plants that are critical to the nation's infrastructure. The security assessment of NCMS is thus critical to protection from the rising danger of cyber risks.

8.1 A networked cellular manufacturing system

A networked cellular manufacturing system is a congregation of independent cellular manufacturing systems that measure and report in real time both local and geographically remote distributed processes. It is a combination of telemetry and data acquisition that enables a user to send commands to distant facilities and collect data from them. Telemetry is a technique used in transmitting and receiving data over a medium. Data acquisition is a method of collecting data from the equipment being controlled and monitored. NCMSs are usually networked, with at least one controller, and one or more remote terminal units (RTUs) distributed throughout the plant at necessary control points. An RTU can be compared to an analog or digital converter. It can receive signals from sensors at the control point, convert and send these signals to the controller, and also receive signals back from the controller and convert the commands so they can be understood by the actual control mechanisms. The fundamental components of the NCMS control system are

- Master terminal unit (MTU)
- Communication network (CN)
- Remote terminal unit (RTU)

The supervisory control and monitoring station, also called the master terminal unit, consists of an engineering workstation, human–machine interface, application servers, and communications router. The master terminal unit issues commands to distant facilities, gathers data from them, interacts with other systems in the corporate intranet for administrative purposes, and interfaces with human operators. The MTU has full control on the distributed remote processes. Commands sent from the MTU to distant facilities can be done either manually using a human–machine interface or by automation. The goal of an NCMS project is to enable the RTU to be viewed or controlled remotely, while still maintaining as much security as possible. This can be important for plants that have, for example, operators and troubleshooters off-site or outside the plant's systems. If an employee at the plant reports the failure of a part or system, especially a mission-critical part or system, and for some reason or another cannot access the system or maybe the controls are compromised, another person (outside of the plant) can be called to try to rectify the situation. That is not to say that this external access is a fail-safe; there still needs to be other contingency plans. This is merely a convenience.

A human–machine interface program runs on the MTU computer. This basically consists of a diagram that mimics the whole plant, making it easier to identify with the real system. Every input and output point of the remote systems can be represented graphically with the current configuration parameters being displayed. Configuration parameters such as trip values and limits can be entered onto this interface. This information will be communicated through the network and downloaded onto the operating systems of the corresponding remote locations, which will update all the values. A separate window with a list of alarms set up in the remote station network can also be displayed. The window displays the alarm tag name, description, value, trip point value, time, date, and other important information. Trend graphs can also be displayed. These graphs show the behavior of a certain unit by logging values periodically and displaying it in a graph. If any abnormal behavior of the unit is seen then the appropriate actions can be taken at the right time. The remote sites are known as field sites. The field site basically consists of so-called field instrumentation, which are devices that are connected to the equipment or machines being controlled and monitored by the NCMS. The devices include sensors to monitor certain parameters and actuators for controlling certain modules of the system. Other devices in the field sites are controllers and pulse generators. These devices convert physical parameters to electrical signals that are readable by the remote station equipment. The outputs can be read in either analog or digital form. Generally, voltage outputs have fixed levels, such as 0 to 5V and 0 to 10V. Voltage levels are transmitted when sensors are located close to the controllers and current levels are transmitted when they are located far from the controllers. Digital readings can be used to check if the system has been enabled or disabled, that is, in operation or out of operation. Actuators help in sending out commands to the equipment, that is, turn on and off the equipment.

The field instrumentation we just described is interfaced with the controller called the remote terminal unit (RTU) or programmable logic controller (PLC). Both of these basically consist of a computer controller that can be used for process manipulation at the remote site. They are interfaced with the communication system connected to the MTU. The PLC has very good programmability features, whereas RTUs have better interfaces to the communication lines. The advancement in this area is the merging of PLCs and RTUs to exploit both the features. Hence, the overall function of this architecture is that the MTU communicates with one or more remote RTUs by sending requests for information that those RTUs gather from devices, or instructions to take an action such as open and close valves, and turn switches on and off.

An intelligent electronic device (IED) is a protective relay that communicates with the RTU. A number of IEDs can be connected to the RTU. They are all polled and data is collected. IEDs also have a direct interface to control and monitor sensory equipment. IEDs have local programming that allows it to act without commands from the control center. This makes the RTU more automated and even the amount of communication with the MTU is reduced.

Communication media used between MTUs and RTUs vary from wired networks, such as public switched telephone networks, to using wireless or radio networks. The MTU and the administrative systems are connected in a LAN (local area network). In the communication medium between the MTU and RTU, the most commonly used protocols are the Distributed Network Protocol (DNP3) and Modbus. DNP3 is an open standard and a relatively new protocol. The older systems use the Modbus protocol. DNP3 and Modbus have been adopted by a number of vendors that support the NCMS. Both the DNP3 and Modbus protocols have been extended to be carried over TCP/IP. Also connected to the control system discussed earlier is an enterprise network. This connectivity provides decision makers with access to real-time information, and allows engineers to monitor and control the control system.

The architecture described above has a number of vulnerabilities. The MTU and RTUs are connected via the Internet, public switched telephone network (PSTN), cable, or wireless. The most common security issue in all the aforementioned communication networks is eavesdropping. Wireless networks and the Internet are prone to replay attacks, denial of service attacks, and so forth. Outside vendors, consumers, and business partners can carry out attacks on this architecture since they are connected to the enterprise network through an Internet connection. Hence, these entities have indirect access to the MTU since the enterprise network is connected to the control system. Remote stations have a communication interface that allows field operators to communicate via wireless protocol or remote modem to perform maintenance operations. These operations are done using handheld devices. When an unauthorized person gets access to this handheld device, they could cause harm to the system. There are several more security issues in this architecture.

8.2 Cyclical redundancy check for a remote terminal unit mode

A cyclical redundancy check (CRC) is the error checking used for the RTU mode. In the RTU mode, messages include an error-checking field that is based on a CRC method. The CRC field checks the contents of the entire message. It is applied regardless of any parity check method used for the individual characters of the message.

- Let E be the *error mask* caused by the channel, that is, a bit string indicating which bits got corrupted.
- Let F = M;R (the bit frame sent by the sender).
- The receiver receives F + E, which equals F XOR E (recall modulo 2).
- If there were no corrupted bits, E = 0. Otherwise, E ≠ 0.

The CRC field is two bytes, containing a 16-bit binary value. The CRC value is calculated by the transmitting device, which appends the CRC to the message. The receiving device recalculates a CRC during receipt of the message, then compares the calculated value to the actual value it received in the CRC field. If the two values are not equal, an error results.

Example 8.1

- Receiver receives F + E for some F and some E.
- If F = M;R = 110011 and E = 101, then the bit string received by the receiver is (F + E), that is,

$$F + E = 110110$$

- The first and third bits from right to left got corrupted.
- Of course, the receiver does not know E.

The CRC is started by first preloading a 16-bit register to all 1s. Then a process begins of applying successive 8-bit bytes of the message to the current contents of the register. Only the 8 bits of data in each character are used for generating the CRC. Start and stop bits, and the parity bit, do not apply to the CRC. During the generation of the CRC, each 8-bit character is exclusive ORed with the register contents. Then the result is shifted in the direction of the least significant bit (LSB), with a zero filled into the most significant bit (MSB) position. The LSB is extracted and examined. If the LSB was a 1, the register is then exclusive ORed with a preset, fixed value. If the LSB was a 0, no exclusive OR takes place.

- The receiver performs the division with the bits it received.
- The receiver performs (F + E)/G and checks the remainder.
- If the remainder is zero, it assumes E = 0 and no corruption occurred.
- If the remainder is nonzero, it assumes E ≠ 0 and corruption occurred.

This process is repeated until eight shifts have been performed. After the last (eighth) shift, the next 8-bit byte is exclusive ORed with the register's current value, and the process repeats for eight more shifts as described earlier. The final contents of the register, after all the bytes of the message have been applied, is the CRC value. When the CRC is appended to the message, the low-order byte is appended first, followed by the high-order byte. In ladder logic, the CKSM function calculates a CRC from the message contents.

- Notice that (F + E)/G = F/G + E/G.
- Also, since F = M;R, then F/G has a zero remainder.
- Thus, (F + E)/G has a nonzero remainder if and only if E/G has a nonzero remainder.
- Thus, to show if corruption errors can be detected, we do not concern ourselves with F (the original message), we focus on the error mask E.
- Next, for different types of error masks E, we see if the protocol can detect the corruption (i.e., if a nonzero remainder is obtained).

8.2.1 Detecting single-bit errors

CRC can detect single-bit errors. Observe that the requirement on the divisor is that it has two or more nonzero terms. However, the requirement that x^k and x^0 terms be nonzero is quite natural. If the x^k term is zero the divisor polynomial is not of degree k. If the x^0 term is zero, then the divisor polynomial is divisible by x^n for some $n > 0$, so the low-order bits of the error code are wasted.

- If a single bit got corrupted, then E is of the form 1000…00, that is, E is a 1 followed by some (or no) zeros.
- To detect corruption, E/G must have a nonzero remainder.
- Thus, if G has at least two 1s in it, then E/G has a nonzero remainder, that is, G could be 1000100 (it could have more than two 1s of course, but at least two are necessary).

Example 8.2

A bit stream 10011101 is transmitted using the standard CRC method described in the text. The generator polynomial is $x^3 + 1$. Show the actual bit string transmitted. Suppose the third bit from the left is inverted during transmission. Show that this error is detected at the receiver's end.

Answer

- The frame is 1001, so dividing 10011101000 by 1001, the remainder is 100. So the actual bit string transmitted is 10011101100.
- If the error occurs, the bit string becomes 10111101100, then dividing it by the generator 1001, we can get the remainder 100. As the remainder is not zero, the receiver detects the error and can ask for a retransmission.

8.2.2 *Detecting double-bit errors*

CRC can detect double-bit errors provided the divisor polynomial has a factor with at least three terms. Students observed in class that they had found a counterexample. There is a class of divisor polynomials called "primitive polynomials" that are able to detect double-bit errors quite far apart—far enough apart that they could detect all double-bit errors in messages of practical lengths.

- Two bits get corrupted, so E is of the form 10^j10^k, where 0^j means a string of j zeros, and $j \geq 0$ and $k \geq 0$.
- In this case, we know E/G has a nonzero remainder provided both of the following conditions hold:
  - The rightmost bit of G is a 1.
  - G does not divide the string 10^j1 for any j up to the message size.
- Why? If G does not divide 10^j1, then up to this point in the division we end up with a nonzero remainder.
- All future steps of the division bring down a 0 bit from E, which will get "matched" against the rightmost bit of G, which is a 1, and thus, a nonzero remainder is obtained.
- Consider the string 10^j1, which as a polynomial is of the form $x^{(j-1)} + 1$.
- The smallest value of $(j - 1)$ such that a polynomial divides $x^{(j-1)} + 1$ is known as the *order* or *exponent* of the polynomial.
- The polynomials with the highest order are known as *primitive polynomials.*
- For polynomials with 0 1 coefficients, primitive polynomials of degree r have an order of $2^r - 1$.
- Thus, if G has a primitive polynomial of degree r as a factor, it can catch all double-bit errors in messages up to size 2^r bits!

Suppose we want to be able to detect all error patterns with two errors. That error pattern may be written as $x^i + x^j = x^i(1 + x^{j-i})$, for some i and j > i. If g(x) does not divide this term, then the resulting CRC can detect all double errors.

CRCs are examples of cyclic codes, which have the property that if c is a code word, then any cyclic shift (rotation) of c is another valid code word. Hence,

$$w(x) = \sum_{i=0}^{n-1} w_i x^i \tag{8.1}$$

we find that one can represent the polynomial corresponding to one cyclic left shift of w as

$$w^{(1)}(x) = w_{n-1} + w_0 x + w_1 x^2 + \ldots w_{n-2} x^{n-1} \tag{8.2}$$

$$= xw(x) + (1 + x^n)w_{n-1} \tag{8.3}$$

Now, because $w^{(1)}(x)$ must also be a valid code word, it must be a multiple of g(x), which means that g(x) must divide $1 + x^n$.

Note that $1 + x^n$ corresponds to a double error pattern; what this observation implies is that the CRC scheme using cyclic code polynomials can detect the errors we want to detect (such as all double-bit errors) as long as g(x) is picked so that the smallest n for which $1 + x^n$ is a multiple of g(x) is quite large. For example, in practice, a common 16-bit CRC has a g(x) for which the smallest such value of n is $2^{15} - 1 = 32767$, which means that it is quite effective for all messages of length smaller than that.

- All double-bit errors $E(x) = x^i + x^j = x^i(1 + x^{j-i})$, if G(x) is a primitive with a degree of c and the length of code word is less than $2^c - 1$. It implies that $j - i <= 2^c - 1$.
- All double-bit (adjacent-bit) errors will be detected.

8.2.3 *Detecting odd numbers of bits*

Now suppose we want to detect all odd numbers of errors. If (1 + x) is a factor of g(x), then g(x) must have an even number of terms. The reason is that any polynomial with coefficients in F2 of the form (1 + x)h(x) must evaluate to 0 when we set x to 1. If we expand (1 + x)h(x), if the answer must be 0 when x = 1, the expansion must have an even number of terms. Therefore, if we make 1 + x a factor of g(x), the resulting CRC will be able to detect all error patterns with an odd number of errors. Note, however, that the converse statement is not true: a CRC may be able to detect an odd number of errors even when its g(x) is not a multiple of (1 + x). But all CRCs used in practice do have (1 + x) as a factor because it is the simplest way to achieve this goal.

- Assume E contains an odd number of corruptions (an odd number of 1s in E).
- Think of E as the coefficients of a polynomial, that is, if E = 1011, then

$$E = X^3 + X + 1$$

- Claim: If E has an odd number of 1s, then X + 1 is not a factor of E.
- Assume X + 1 is a factor of E. Then, E = (X + 1)*P for some polynomial P.
- If we evaluate E at 1, that is, if we assign 1 to X then

$$E(1) = (1 + 1)*P(1) = 0*P(1) = 0 \text{ (recall modulo-2 arithmetic)}$$

- However, since E has an odd number of terms, then E(1) = 1, due to modulo-2 addition. Hence, X + 1 cannot be a factor of E.
- Therefore, if we choose G such that X + 1 is a factor of G, then E/G has a nonzero remainder because E does not have X + 1 as a factor.

CRC can detect any odd number of errors as long as C contains the factor $(x + 1)$. Suppose the error polynomial, E, is divisible by C. (That is E mod C = 0.) Then for some polynomial, D, E = $(x + 1)$D. Now we use the observation that if two functions are equal they must evaluate to the same thing given the same argument. So E(1) = (1 + 1)D(1) = 0, which can only be true if E has an even number of terms. Thus, C will detect any odd number of errors.

8.2.4 Detecting burst errors

CRC can detect any burst error as long as the burst is of less than k bits. Such an error polynomial is expressible as E = Dx^n where the degree of D is less than k. Since from part a above, x^n is not a factor of the divisor polynomial, C, E mod C = 0 implies D mod C = 0, which requires that the degree of D be at least that of C. Hence all burst errors of less than k bits will be detected.

Error burst of k bits, $2 \leq k \leq n$

- In this case, $E = 1b^i10^j$, where each of the i bits in b^i can be either 0 or 1, $i \leq n - 2$, and $0 \leq j$.
- For example, E = 101100100000.

- $1(b^i)1 = 1011001$ and $0^j = 00000$.
- Let the rightmost bit of G be 1.
- Since G has n + 1 bits, it is longer than $1b^i1$, and ($1b^i1$/G) has a nonzero remainder.
- Since the remainder after the division steps up to $1b^i1$ is nonzero, then the remainder at the end of the division is nonzero (the next bit dropped is always a 0, which is matched against the 1 in G).

- Error burst of n + 1 bits
    - E and G have the same number of bits.
    - There is no guarantee that the remainder is nonzero, since if by chance E = G the remainder is obviously zero.
    - The probability of the remainder being zero given that the burst size is n + 1 (and thus the message is accepted incorrectly) is $1/2^{n-1}$.
    - Why $1/2^{n-1}$ and not $1/2^{n+1}$? The first and last bits do not count, since they are always fixed to be 1.
- Error burst greater than n + 1 bits
    - In this case, the remainder is zero with probability $1/2^n$.
    - 2^n possible remainders, you get one randomly.

Another guideline used by some CRC schemes in practice is the ability to detect burst errors. Let us define a burst error pattern of length b as a sequence of bits $1\varepsilon_{b-2}\varepsilon_{b-3}\ldots\varepsilon_1 1$: that is, the number of bits is b, the first and last bits are both 1, and the bits εi in the middle could be either 0 or 1. The minimum burst length is 2, corresponding to the pattern "11." Suppose we would like our CRC to detect all such error patterns, where

$$e(x) = X^s \left(1 \cdot X^{b-1} + \sum_{i=1}^{b-2} \varepsilon_i \cdot X^i + 1 \right)$$

This polynomial represents a burst error pattern of size b starting s bits to the left from the end of the packet. If we pick g(x) to be a polynomial of degree b, and if g(x) does not have x as a factor, then any error pattern of length ≤b is guaranteed to be detected, because g(x) will not divide a polynomial of degree smaller than its own. Moreover, there is exactly one error pattern of length b + 1 (corresponding to the case when the burst error pattern matches the coefficients of g(x) itself) that will not be detected. All other error patterns of length b + 1 will be detected by this CRC. In fact, such a CRC is quite good at detecting longer burst errors as well, though it cannot detect all of them.

Example 8.3

The 16-bit standard CRC ($n = 16, x^{16} + x^{15} + x^2 + 1$), catches:

- All single-bit errors
- All double-bit errors
- All odd number of bit errors
- All bursts of length at most 16
- 99.9969% of length 17
- 99.9985% of length greater than 17

8.3 Hazard and operability study of cellular manufacturing: Process safety management

Hazard and operability (HAZOP) analysis is a structured and systematic technique for system examination and risk management. In particular, HAZOP is often used as a technique for identifying potential hazards in a system and identifying operability problems likely to lead to nonconforming products. HAZOP is based on a theory that assumes risk events are caused by deviations from design or operating intentions. Identification of such deviations is facilitated by using sets of "guide words" as a systematic list of deviation perspectives. This approach is the unique feature of HAZOP.

HAZOP methodology helps stimulate the imagination of team members when exploring potential deviations.

A HAZOP study identifies hazards and operability problems in a process plant. It is a tool for the identification of hazards due to process parameter deviations. The concept involves investigating how the plant might deviate from the design intent. HAZOP is based on the principle that several experts with different backgrounds can interact and identify more problems when working together than when working separately and combining their results. Although the HAZOP study was developed to supplement experience-based practices when a new design or technology is involved, its use has expanded to almost all phases of a plant's life. The "guide word" HAZOP is the most well known of the HAZOPs; however, several specializations of this basic method have been developed.

8.3.1 The HAZOP concept

The HAZOP concept is to review the plant in a series of meetings, during which a multidisciplinary team methodically brainstorms the plant design, following the structure provided by the guide words and the team leader's experience.

The primary advantage of this brainstorming is that it stimulates creativity and generates ideas. This creativity results from the interaction of the team members and their diverse backgrounds. Consequently, the process requires that all team members participate (quantity breeds quality in this case), and team members must refrain from criticizing each other to the point that members hesitate to suggest ideas.

The team focuses on specific points of the design (called "study nodes") one at a time. At each of these study nodes, deviations in the process parameters are examined using the guide words. The guide words are used to ensure that the design is explored in every conceivable way. Thus, the team must identify a fairly large number of deviations, each of which must then be considered so that their potential causes and consequences can be identified. The success or failure of a HAZOP study depends on several factors, to name a few:

- The completeness and accuracy of drawings and other data used as a basis for the study
- The technical skills and insights of the team
- The ability of the team to use the approach as an aid to its imagination in visualizing deviations, causes, and consequences
- The ability of the team to concentrate on the more serious hazards that are identified

In the process of identifying problems during a HAZOP study, if a solution becomes apparent, it is recorded as part of the HAZOP result; however, care was taken to avoid trying to find solutions that are not so apparent, because the prime objective for the HAZOP is problem identification.

8.3.2 *Terms used in a HAZOP study*

The HAZOP process is systematic and it is helpful to define the terms that are used in the study:

Study nodes—The locations (on piping and instrumentation drawings and procedures) at which the process parameters are investigated for deviations.

Intention—The intention defines how the plant is expected to operate in the absence of deviations at the study nodes. This can take a number of forms and can either be descriptive or diagrammatic, for example, flow sheets, line diagrams, and piping and instrumentation diagrams (P&IDs).

Deviations—These are departures from the intention that are discovered by systematically applying the guide words (e.g., "more pressure").

Causes—These are the reasons why deviations might occur. Once a deviation has been shown to have a credible cause, it can be treated as a meaningful deviation. These causes can be, for example, hardware failures, human errors, an unanticipated process

Table 8.1 HAZOP Guide Words and Meanings

Guide words	Meaning
No	Negation of the design intent
Less	Quantitative decrease
More	Quantitative increase
Part Of	Qualitative decrease
As Well As	Qualitative increase
Reverse	Logical opposite of the intent
Other Than	Complete substitution

state (e.g., change of composition), or external disruptions (e.g., loss of power).

Consequences—These are the results of the deviations should they occur (e.g., release of toxic materials). Trivial consequences, relative to the study objective, are dropped.

Guide words—These are simple words that are used to qualify or quantify the intention in order to guide and stimulate the brainstorming process and so discover deviations. The guide words shown in Table 8.1 are the ones most often used in a HAZOP study; some organizations have made this list specific to their operations, to guide the team more quickly to the areas where they have found problems previously. Each guide word is applied to the process variables at the point in the plant (study node) that is being examined. These guide words are applicable to both the more general parameters (e.g., react, transfer) and the more specific parameters (e.g., pressure, temperature).

HAZOP guide words and meanings are summarized in Table 8.1. With the general parameters, meaningful deviations are usually generated for each guide word. Moreover, it is not unusual to have more than one deviation from the application of one guide word. For example, "more reaction" could mean either that a reaction takes place at a faster rate or that a greater quantity of product results. With the specific parameters, some modification of the guide words may be necessary. In addition, it is not unusual to find that some potential deviations are eliminated by physical limitation. For example, if the design intention of a pressure or temperature is being considered, the guide words "more" or "less" may be the only possibilities.

Finally, when dealing with a design intention involving a complex set of interrelated plant parameters (e.g., temperatures, reaction rates, composition, or pressure), it may be better to apply the whole sequence of guide words to each parameter individually than to apply each guide word across all of the parameters as a group. Also, when applying the guide words to a sentence it may be more useful to apply the sequence of guide words to each word or phrase separately, starting with the key part that describes the activity (usually the verbs or adverbs). These parts of the sentence are usually related to some impact on the process parameters.

8.3.3 HAZOP methodology

The aforementioned concepts are put into practice in the following steps:

1. Define the purpose, objectives, and scope of the study
2. Select the team
3. Prepare for the study
4. Carry out the team review
5. Record the results

It is important to recognize that some of these steps can take place at the same time. For example, the team reviews the design, records the findings, and follows up on the findings continuously. For process safety assessment, a safety instrumented system (SIS) consists of an engineered set of hardware and software controls that are especially used on critical process systems. A critical process system can be identified as one that, once running and an operational problem occurs, may need to be put into a "safe state" to avoid adverse safety, health, and environmental (SH&E) consequences. The SIS has become more important in recent years due to accidents in the process industries. These incidents led engineering organizations to develop best practices and standards. These standards suggested that a separate system needed to be implemented for safety functions away from the basic process control system (BPCS). A separate SIS minimized the risk of common cause failures. This separation has become an industry standard.

Example 8.4

Figure 8.1 shows a chemical reactor as an example of a cellular manufacturing system (CMS) with a safety instrumented system (SIS). FS is the flow switch, which is the logic needed for a single measurement to activate the SIS.

When the switch detects an airflow measurement below the minimum airflow limit, the switch will shut the fuel. To prevent unnecessary shutdowns, we can require that the flow remain below its minimum for a specified time (e.g., 1 second) so that a very short-term fluctuation, that is quickly corrected, does not activate the SIS. Because the SIS is extremely simple, involving only one sensor and a switch, it can be completely documented on the drawing. Note that an extra valve has been added because the control valve might not provide tight shutoff. A separate valve with tight shutoff is provided. Also, shutting two valves in a series provides higher reliability.

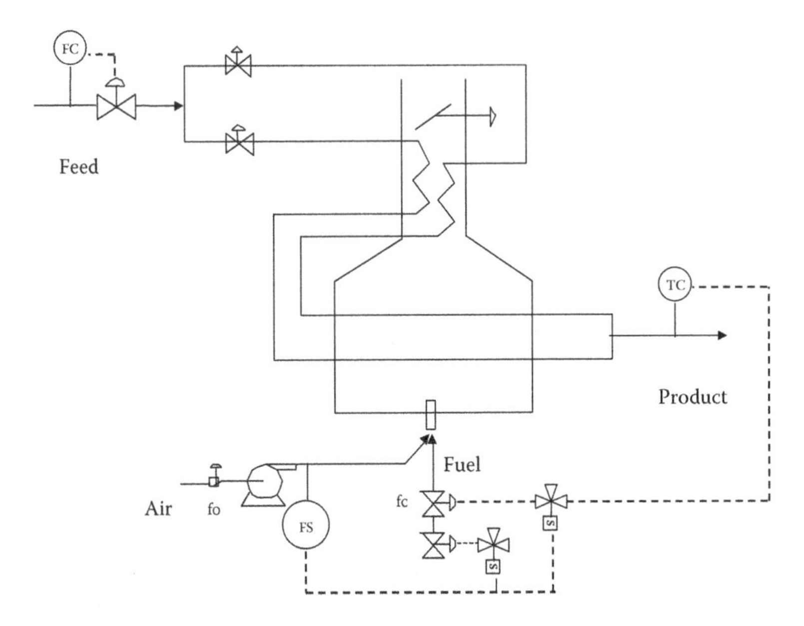

Figure 8.1 A chemical reactor as an example of a cellular manufacturing system (CMS) with a safety instrumented system (SIS).

8.3.4 Risk and risk matrix in a HAZOP study

In the context of process safety management, *risk* is defined in terms of the likelihood and consequences of incidents that could expose plant personnel, property, assets, process, and environment to the harmful effects of a hazard. According to the Centre for Chemical Process Safety of the American Institute of Chemical Engineers, hazards are potential sources of harm, including chemical or physical conditions or characteristics that can damage people, property, or the environment. Incident likelihood encompasses frequency and probability; consequences refer to outcomes and impacts.

Risk ranking uses a matrix that has ranges of consequence and likelihood as the axes. A typical risk matrix is a 4 × 4 grid. A risk matrix is a matrix that is used during risk assessment to define the various levels of risk as the product of the harm probability categories and harm severity categories. This is a simple mechanism to increase the visibility of risks and assist in management decision making.

8.3.5 Design of a risk matrix in a HAZOP study

Although there are many risk matrices that have been developed, effective risk-ranking tools in day-to-day operations, such as during HAZOP studies, are very limited. A layers of protection analysis (LOPA) approach is one of them. It is simple to implement and easy for most HAZOP participants to understand. LOPA is a powerful analytical tool for assessing the adequacy of protection layers used to mitigate process risk. LOPA builds upon well-known process hazards analysis techniques, applying semiquantitative measures to the evaluation of the frequency of potential incidents and the probability of failure of the protection layers.

8.3.6 *Layers of protection in the cellular manufacturing process*

A HAZOP study deals with the identification of hazards due to process parameter deviations. When a failure occurs due to deviations, it may take the process outside of its normal operating ranges. In general, there are several layers of protection measures in a plant in response to a process deviation. The basic process controls, alarms, safety valves, operator supervision, and so on, are the typical protection measures against any harmful consequences due to deviation of process parameters as shown next:

- Process equipment is designed for process operating limits.
- Basic process controls, alarms, and operators are adjusted to process deviations.
- Presence of critical alarms along with speedy response of operators.
- Safety interlock system/emergency shutdown at operating limits.
- Relief systems that activate at equipment design limits.
- Mitigation systems that contain the effects of an incident.
- Plant emergency response to control the effects of incidents (on-site control arrangement).
- Emergency response to protect the public from the effects of an incident (off-site control arrangement).

The basic process control system (BPCS) is the lowest layer of protection and is responsible for the normal operation of the plant. If this system fails or is incapable of maintaining control, then the second layer (operator intervention) attempts to resolve the problem. If the operator cannot maintain control within certain limits, then the safety system layer must attempt to bring the plant to a safe condition. For this hierarchy to be effective it is critical that each layer be independent or separate. This means that the two layers (BPCS and SIS) must not contain common components that in the event of a failure would actually prevent the SIS layer from protecting the facility and people when the BPCS layer experiences a problem. There would be little value in using a single transmitter connected to the BPCS and the safety system. If this transmitter caused the failure by giving false information to the BPCS it could not be relied upon to give accurate information to the safety system. This idea of eliminating common cause faults has led to many discussions about separation. For this reason the end devices and the logic solvers in each system need to be separate from the BPCS. Therefore, responsible designers and governing bodies have made standards that enforce this separation.

8.3.7 Likelihood and consequence ranges

In the LOPA approach, the highest likelihood range (level 4) is defined by the likelihood of the initiating event, for example, human error or control failure. Then for each level of existing protection measure, the likelihood range is reduced by one level. This approach assumes that each level of protection has a similar failure probability, which is generally acceptable for rough risk screening such as HAZOP risk ranking. Some failures have fairly well-defined frequencies of occurrences and can be directly used. For example, catastrophic failure of a pressure vessel has a frequency in the range of 10^{-5} per year and thus by itself would be considered as a level 1 likelihood. Similar likelihood levels can be defined for other common equipment failures, pipe leaks and ruptures. The likelihood ranges can be used in conjunction with typical consequence ranges to people, property, and environment.

One option is to avoid using quantitative frequencies or probabilities for the likelihood ranges and use the LOPA approach as shown in Table 8.2. Tables 8.2 and 8.3 show the likelihood ranges and consequence ranges, respectively. In the likelihood range, the highest chances of occurrences have been given to leaking scenarios and so forth associated with the least protection measures in a plant. Similarly, catastrophic failure of tanks or process vessels have the least chances of occurrence and are also associated with various protection measures against failures. In the consequence ranges, major aspects, namely, human injury or

***Table* 8.2** Likelihood Ranges Based on Levels of Protection

Likelihood range	Qualitative frequency criteria: Typical scenarios
1	• Three levels of protection • Tank/process vessel failures
2	• Two levels of protection • Full-bore failures of small process lines or fittings
3	• One level of protection • Piping leaks
4	• Initiating event/human error • Hose leaks/rupture

Table 8.3 Typical Consequence Ranges

Consequence range	Qualitative safety consequence criteria
1	• Injuries requiring first-aid only • Product quality affected and damage value range: Rs. 10,000–1 lakh • Contained release with local environmental impact and pollution problem
2	• Injuries requiring a physician's care • Damage value range: Rs. 1–10 lakh • Uncontained release with potential for minor environmental impact • Chances of fire and explosion
3	• Severe injuries or potential for a fatality • Damage value range: Rs. 10 lakh–1 Crore • Uncontained release with potential for moderate environmental impact
4	• Multiple life-threatening injuries and/or fatality • Damage value range: More than Rs. 1 Crore • Uncontained release with potential for major environmental impact

fatality, environmental impact, production loss, and product quality have been considered.

As shown in Table 8.4, HAZOP is a power tool that is used to identify safety instrumented functions (SIFs). The SIS is a set of components such as sensors, logic solvers, and final control elements arranged for the purpose of taking the process to a safe state when predetermined conditions are violated. Another view is that it is a collection of SIFs.

An SIF is a loop composed of one or more transmitters and one or more valves linked together for the purpose of preventing hazards. Each SIF is rated at a safety integrity level (SIL) based upon the consequence and frequency of occurrence. In the past, process plants used different methods to define the SIL of their plant. Often SIL 3 was considered a "worse case" and plants were designed around this rating. This led to overengineering and expensive systems. Current standards require that each SIF in an SIS be considered separately. This means that there are no "SIL 3 plants." There are process plants that may be a combination of SIL 3, SIL 2, SIL 1, and SIL 0 SIFs. After each SIF has been assigned an SIL level, a risk reduction factor (RRF) must be determined. The RRF is the reduction in risk that has to be achieved to meet the tolerable risk for a specific situation.

Table 8.4 HAZOP for a Chemical Reactor (Partial)

Unit: Chemical Reactor

Node: Air pipe after compressor and valve Parameter: Pressure

Guide word	Deviation	Cause	Consequence	Action
Lower	Low pressure in the fuel pipe node	Stoppage of power to motor or turbine turning the compressor	Uncombusted fuel in the fire box, danger of explosion Uncombusted fuel, wasted fuel	SIS based on the rotation of motor shaft
		Break of coupling between motor and compressor	Uncombusted fuel in the fire box, danger of explosion	SIS based on rotation of compressor shaft
		Failure of compressor, e.g., breakage of blades	Uncombusted fuel, wasted fuel Danger from flying metal	
		Closure of air valve due to failure	Uncombusted fuel in the fire box, danger of explosion	Fail open valve
		Any of the above	Uncombusted fuel, wasted fuel	SIS that measures the flow of air after the pipe and activates the shutdown of the flow if too low
		Closure of air valve due to poor decision by operator		Air flow controller with ratio to fuel flow

8.3.8 Risk ranking matrix and acceptability values

Risk assessment is an effective means of identifying process safety risks and determining the most cost-effective means to reduce risk. Many organizations recognize the need for risk assessment, but most do not have the tools, experience, and resources to assess risk quantitatively. Therefore, these organizations use qualitative or semiquantitative risk assessment tools, such as risk ranking. Use the risk-ranking matrix by plotting the probability of the occurrence versus the consequences of the occurrence. Although risk matrices are easy to use, unless they are properly designed, they can create liability issues and give a false sense of security. An effective risk-ranking matrix should have the following characteristics:

- Be simple to use and understand
- Not require extensive knowledge of quantitative risk analysis to use
- Have clear guidance on applicability
- Have consistent likelihood ranges that cover the full spectrum of potential scenarios
- Have detailed descriptions of the consequences of concern for each consequence range
- Have clearly defined tolerable and intolerable risk levels
- Show how scenarios that are at an intolerable risk level can be mitigated to a tolerable risk level on the matrix
- Provide clear guidance on what action is necessary to mitigate scenarios with intolerable risk levels

Risk ranking is a common methodology for making risk-based decisions without conducting quantitative risk analysis. Based on the likelihood and consequence ranges, risk ranking has been defined with suitable acceptability criteria as shown in Table 8.5.

For a CMS, process hazards should be defined, and process risk evaluation criteria and product risk acceptability values should be established at the beginning of CMS development. As CMS development proceeds, risk assessments should identify the risks to be minimized or controlled through CMS design. At the end of CMS process development, companies should have a documented understanding of CMS risks and their associated

Table 8.5 Risk Rank Value and Acceptability Criteria

Risk rank value	Explanation
1–2	Acceptable as it is
3–4	Acceptable with controls
5–9	Undesirable; mitigation required to reduce risk
>9	Unacceptable

controls. This information should be communicated during technology transfer to the commercial manufacturing operation or contractor.

8.4 Automation

Automation is the use of machines, control systems, and information technologies to optimize productivity in the production of goods and delivery of services. It is the transfer of human intelligence to automated machinery so that machines are able to stop, start, load, and unload automatically. In many cases, machines can also be designed to detect the production of a defective part, stop themselves, and signal for help. The correct incentive for applying automation is to increase productivity or quality beyond that possible with current human labor levels so as to realize economies of scale, or realize predictable quality levels. This frees operators for other value-added work.

In the scope of industrialization, automation is a step beyond mechanization. Whereas mechanization provides human operators with machinery to assist them with the muscular requirements of work, automation greatly decreases the need for human sensory and mental requirements while increasing load capacity, speed, and repeatability. Automation plays an increasingly important role in the world economy and in daily experience. This concept has also been known as "automation with a human touch" and jidoka, and it was pioneered by Sakichi Toyoda in the early 1900s when he invented automatic looms that stopped instantly when any thread broke. This enabled one operator to manage many machines without the risk of producing vast amounts of defective cloth. This technique is closely linked to mistake-proofing, a risk mitigation technique.

Bibliography

Black, J. T. 1991. *The Design of the Factory with a Future*. New York: McGraw-Hill.

Black, J. T. 2000. Lean manufacturing implementation. In *Innovations in Competitive Manufacturing*, edited by Paul M. Swamidass, pp. 177–186. Boston: Kluwer Academic.

Brandon, J. 1996. *Cellular Manufacturing: Integrating Technology and Management*. Somerset, England: Research Studies Press LTD.

Burbidge, J. L. 1978. *The Principles of Production Control*. England: MacDonald and Evans.

Feld, W. M. 2001. *Lean Manufacturing: Tools, Techniques, and How to Use Them*. Boca Raton, FL: St. Lucie Press; Alexandria, VA: APICS.

Flanders, R. E. 1925. Design manufacture and production control of a standard machine. *Transactions of ASME*, 46, 691–738.

Hyer, N., and Wemmerlov, U. 2002. *Reorganizing the Factory: Competing through Cellular Manufacturing*. Portland, OR: Productivity Press.

Irani, S. 1999. *Handbook of Cellular Manufacturing Systems*. New York: John Wiley & Sons.
Mitrofanov, S. P. 1966. *The Scientific Principles of Group Technology*. Boston: National Lending Library for Science and Technology.
Nolan, D. P. 1994. *Application of HAZOP and What-If Safety Reviews to the Petroleum, Petrochemical and Chemical Industries*. Norwich, NY: William Andrew; Park Ridge, NJ: Noyes.
Peterson, W. W., and Brown, D. T. 1961. Cyclic codes for error detection. In *Proceedings of the IRE*, January, pp. 228–235.
Singh, N., and Rajamani, D. 1996. *Cellular Manufacturing Systems Design, Planning and Control*. London, UK: Chapman & Hall.
Swamdimass, P. M., and Darlow, N. R. 2000. Manufacturing strategy. In *Innovations in Competitive Manufacturing*, edited by P. M. Swamidass, pp. 17–24. Boston: Kluwer Academic.
Swann, C. D., and Preston, M. L. 1995. Twenty-five years of HAZOPs. *Journal of Loss Prevention in the Process Industries*, 8(6), 349–353.
Tanenbaum, A. S. 1988. *Computer Networks*, 2nd ed. Englewood Cliffs, NJ: Prentice Hall.
Wang, J. X. 2002. *What Every Engineer Should Know about Decision Making under Uncertainty*. Boca Raton, FL: CRC Press.
Wang, J. X. 2005. *Engineering Robust Designs with Six Sigma*. Upper Saddle River, NJ: Prentice Hall.
Wang, J. X. 2010. *Lean Manufacturing: Business Bottom-Line Based*. Boca Raton, FL: CRC Press.
Wang, J. X. 2012. *Green Electronics Manufacturing: Creating Environmental Sensible Products*. Boca Raton, FL: CRC Press.
Wang, J. X., and Roush, M. L. 2000. *What Every Engineer Should Know about Risk Engineering and Management*. Boca Raton, FL: CRC Press.
Womack, J. P., Jones, D. T., and Roos, D. 2001. *The Machine That Changed the World*. New York: Harper Perennial.

Epilogue: Cellular manufacturing leadership—Imagination and business growth

Genius is 1% inspiration, 99% perspiration

—Thomas Edison

Cellular manufacturing is about learning and growing the entire team to reduce incidental work and increase output at a safe and consistent pace. Cellular manufacturing focuses on the reduction of the nine deadly wastes also referred to as *muda*:

1. Overproduction
2. Waiting
3. Transportation
4. Conveyance
5. Overprocessing
6. Incorrect processing
7. Excess inventory
8. Unnecessary movement
9. Defects

The objective is to improve profitability and long-term growth and stability. In cellular manufacturing, a change in one component of a product (the heating element of a coffeemaker, for instance) has relatively little influence on the performance of other parts or the system as a whole.

In cellular manufacturing, the components are highly interdependent, and the result is nonlinear behavior: A minor change in one part can cause an unexpectedly huge difference in the functioning of the overall system. With semiconductors, for example, a minuscule impurity (just 10 parts in a billion) can dramatically alter silicon's resistance by a factor of more than 10,000. Generally speaking, traditional manufacturing may make R&D more predictable, but traditional manufacturing processes tend to result in incremental product improvements instead of important advances. Cellular manufacturing systems, on the other hand, are riskier to work with, but they are more likely to lead to breakthroughs.

Remember what Thomas Edison said, "If we all did the things we are capable of doing, we would literally astound ourselves."

Index

C

S

T

For Product Safety Concerns and Information please contact our EU representative GPSR@taylorandfrancis.com Taylor & Francis Verlag GmbH, Kaufingerstraße 24, 80331 München, Germany

Printed and bound by CPI Group (UK) Ltd, Croydon, CR0 4YY

08/05/2025

01864387-0003